SpringerBriefs in Applied Sciences and Technology

Manufacturing and Surface Engineering

Series Editor

Joao Paulo Davim, Department of Mechanical Engineering, University of Aveiro, Aveiro, Portugal

This series fosters information exchange and discussion on all aspects of manufacturing and surface engineering for modern industry. This series focuses on manufacturing with emphasis in machining and forming technologies, including traditional machining (turning, milling, drilling, etc.), non-traditional machining (EDM, USM, LAM, etc.), abrasive machining, hard part machining, high speed machining, high efficiency machining, micromachining, internet-based machining, metal casting, joining, powder metallurgy, extrusion, forging, rolling, drawing, sheet metal forming, microforming, hydroforming, thermoforming, incremental forming, plastics/composites processing, ceramic processing, hybrid processes (thermal, plasma, chemical and electrical energy assisted methods), etc. The manufacturability of all materials will be considered, including metals, polymers, ceramics, composites, biomaterials, nanomaterials, etc. The series covers the full range of surface engineering aspects such as surface metrology, surface integrity, contact mechanics, friction and wear, lubrication and lubricants, coatings an surface treatments, multiscale tribology including biomedical systems and manufacturing processes. Moreover, the series covers the computational methods and optimization techniques applied in manufacturing and surface engineering. Contributions to this book series are welcome on all subjects of manufacturing and surface engineering. Especially welcome are books that pioneer new research directions, raise new questions and new possibilities, or examine old problems from a new angle. To submit a proposal or request further information, please contact Dr. Mayra Castro, Publishing Editor Applied Sciences, via mayra.castro@springer.com or Professor J. Paulo Davim, Book Series Editor, via pdavim@ua.pt

More information about this subseries at http://www.springer.com/series/10623

Kaushik Kumar · Divya Zindani ·
J. Paulo Davim

Design Thinking to Digital Thinking

Kaushik Kumar
Department of Mechanical Engineering
Birla Institute of Technology
Ranchi, Jharkhand, India

Divya Zindani
Department of Mechanical Engineering
National Institute of Technology
Silchar, Assam, India

J. Paulo Davim
Department of Mechanical Engineering
University of Aveiro
Aveiro, Portugal

ISSN 2191-530X ISSN 2191-5318 (electronic)
SpringerBriefs in Applied Sciences and Technology
ISSN 2365-8223 ISSN 2365-8231 (electronic)
Manufacturing and Surface Engineering
ISBN 978-3-030-31358-6 ISBN 978-3-030-31359-3 (eBook)
https://doi.org/10.1007/978-3-030-31359-3

This Springer imprint is published by the registered company Springer Nature Switzerland AG
The registered company address is: Gewerbestrasse 11, 6330 Cham, Switzerland

Preface

Design thinking forms a schematic approach to understand and find solution to problems utilizing the same tools traditionally used by developers of commercial and/or industrial products, processes, and environments. In the mid-twentieth century, the development of creative techniques and new design methods provided the foundation of the concept called design thinking as a secured approach toward problem-solving. Authors like John E. Arnold and L. Bruce Archer were the pioneers in discussing the concept in their books *Creative Engineering* (1959) and *Systematic Method for Designers* (1965), respectively. The technique was designed to provide solution giving novel functionality and higher levels of performance at a lower production costs which would ultimately increase saleability. Accordingly, the steps used were empathize (problem understanding), define (problem definition), ideate (idea generation), prototype, and testing. In the initial phase, these steps were taken one after the other but the gap between the steps and groups involved in the same started providing illogical or erroneous results which created looping of steps and interaction between groups at different stages.

With the introduction of **4th Industrial Revolution** or **I 4.0** resulting in globalization and open market economy, the market has become consumer driven or customer dictated. This has given rise to a situation creating a shift from *Design Thinking* to *Digital Thinking*. So the concept or technique now is amalgamated with Internet, information and communication technologies (ICT), and physical machinery with terms like Internet of things (IoT), industrial Internet of things (IIoT), collaborative robot (cobot), big data, cloud computing, virtual manufacturing, 3D printing, etc., and has marked their presence into all systems. So now the groups involved may not be at a place but can be connected digitally. Moreover, the customer most important participant can oversee the total development and can provide a feedback/suggestion at any stage of the development process. This would ensure more product acceptability, more life expectancy, and hence more sustainability in the business world.

Information science has influenced every aspect of business, society, and human life globally and with tremendous speed. It has brought upon a revolution in the manner mankind communicates and responds, runs business, and even controls others. The presence of technology has clearly indicated in all walks of life, and this effect is following a power law in increase over last couple of years. So the design is also not unaffected by the boom. The present book outlines the paradigm shift from design to digital thinking. This book is primarily intended to provide researchers and students an overview of the current state of affairs dealing with design thinking process and its transition to digital era. Researchers and professionals in an increasing number of fields beyond computer science have been reaping benefits from the digital thinking process. Educators in colleges and universities have begun to promote digital thinking in their respective curricula.

The book, with six chapters, can be divided into two parts. The first part with the first three chapters elaborates *Design Thinking* and next with the last three chapters highlights *Digital Thinking*.

Chapter 1 identifies features of design thinking process and briefs the readers with its importance in enhancing the problem-solving skill. Process of design thinking is a creative and analytical process that provides opportunity to create prototype models, experiment with the prototyped model, and redesign the same after obtaining the feedback. A good design thinker is required to have several characteristics such as creativity and visualization. Hence, the aim of this chapter was to provide a brief summary of the literature on design thinking.

Chapter 2 starts with a brief overview of the history and role of design in engineering realm. The very purpose of design thinking process in engineering realm is to prepare engineering graduates that can design which itself is a complex phenomenon. The chapter presents several design dimensions that corroborate the fact that design is hard to learn and teach. The chapter also outlines the methodologies that aid in suitable learning of design thinking skills. Project-based learning approach has been explored in the chapter which is one of the most-favored models for teaching skills of design thinking. The chapter presents some of the future research directions before making final recommendations.

Chapter 3, the last one in *Design Thinking* part, presents the examination of different tools and methods for design thinking and how these can foster innovation. The key elements of design thinking are viability, feasibility, and desirability. When these overlapping spaces are addressed, the probable chances of innovation increases. One of the major attractions for business communities lies in exploring the possibilities of design thinking methods and tools to underpin the arising innovation within team framework. The identified tools can aid in the generation of ideas and thereby a path to create innovative solutions within a team environment. Guidelines pertaining to the utilization of different design thinking tools and methods have been presented toward the end of the chapter.

Chapter 4, the first one in *Digital Thinking* part, attempts to provide an overview of digital thinking and the challenges associated with its inclusion in engineering education system. Digital thinking has emerged as one of the core capabilities of most of the engineers. Digital thinking encompasses set of skills required to transform real-life challenges to problems. Computer-based knowledge and skills are then used to provide solutions to these problems. Although digital thinking influences everyone, it is one of the challenges to educators to reach the required skill set. An increasing number of arguments are required to be taken into consideration in creation and implementation of university curricula encompassing components of computational thinking.

Chapter 5 presents examples to delineate how digital thinking can be used to address formal and informal education setting. Digital thinking has been considered to be one of the elements of competence and therefore should be made an integral part of child's analytical ability. Therefore, digital thinking must be added to school's learning, and hence, in the present chapter the ways in which digital thinking approach can be integrated into education system have been discussed highlighting challenges associated with appropriately defining digital thinking. The chapter ends with concluding remarks on the future research agenda in the aforementioned subject of discussion.

Design thinking learning resources demands change in the digital era, and therefore, digital thinking approach that is interpretive and user-oriented can be used to fulfill the demands. Chapter 6, the last chapter of the book, primarily highlights the importance of integrating digital thinking into design thinking and also presents opportunities that can aid in the integration of the two most important thinking concepts. The rise of digital age calls for the reconsideration of models for learning and hence the integration of digital thinking into design thinking. On the other hand, design thinking can be employed to engage with the emerging digital tools in order to produce knowledge products whose effect can be felt by generations to come.

First and foremost, we would like to thank God for allowing us and making us capable to believe in our passion and pursue our dreams. Almighty, without your support and blessings this work could not have been done. We would like to thank our ancestors, parents, and relatives for allowing us to follow our ambitions. Our families showed patience and tolerated us for taking yet another challenge which decreases the amount of time we could spend with them. They were our inspiration and motivation. Our efforts will come to a level of satisfaction if the professionals concerned with all the fields get benefitted.

We would also need to thank all the well-wishers, our colleagues, and friends. Their active involvement in the development of this book cannot be expressed in a few words or lines.

We owe a huge thanks to all of our Technical Reviewers, Editorial Advisory Board Members, Book Development Editor, and the team of Springer Nature for their availability for work on this huge project. All of their efforts helped to make this book complete, and we could not have done it without their constant coordination and support.

Last, but definitely not least, we would like to thank all individuals who had taken time out and helped us during the process of writing this book.

Ranchi, India — Kaushik Kumar
Silchar, India — Divya Zindani
Aveiro, Portugal — J. Paulo Davim

Contents

About the Authors

Dr. Kaushik Kumar Associate Professor, Department of Mechanical Engineering, Birla Institute of Technology, Mesra, Ranchi, Jharkhand, 835215, India. E-mail: kkumar@bitmesra.ac.in, kaushik.bit@gmail.com

Kaushik Kumar, B.Tech. (Mechanical Engineering, REC (Now NIT), Warangal), MBA (Marketing, IGNOU), and Ph.D. (Engineering, Jadavpur University), is presently Associate Professor in the Department of Mechanical Engineering, Birla Institute of Technology, Mesra, Ranchi, India. He has 17 years of teaching and research and over 11 years of industrial experience in a manufacturing unit of Global repute. His areas of teaching and research interest are conventional and non-conventional quality management systems, optimization, non-conventional machining, CAD/CAM, rapid prototyping, and composites. He has nine patents, 28 books, 15 edited book volumes, 43 book chapters, 136 international journals, and 21 international and eight national conference publications to his credit. He is on the editorial board and review panel of seven international and one national journals of repute. He has been felicitated with many awards and honors.

Divya Zindani Department of Mechanical Engineering, National Institute of Technology, Silchar, Cachar, Assam, 788010, India. E-mail: divyazindani@gmail.com

Divya Zindani, B.E. (Mechanical Engineering, Rajasthan Technical University, Kota) and M.E. (Design of Mechanical Equipment, BIT Mesra), is presently pursuing Ph.D. in National Institute of Technology, Silchar. He has over 2 years of industrial experience. His areas of interests are optimization, product and process design, CAD/CAM/CAE, rapid prototyping, and material selection. He has one patent, four books, six edited books, 18 book chapters, two SCI journals, seven Scopus indexed international journals, and four international conference publications to his credit.

Prof. J. Paulo Davim Department of Mechanical Engineering, University of Aveiro, Campus Santiago, 3810-193 Aveiro, Portugal. E-mail: pdavim@ua.pt

J. Paulo Davim received his Ph.D. degree in Mechanical Engineering in 1997, M.Sc. degree in Mechanical Engineering (materials and manufacturing processes) in 1991, Mechanical Engineering degree (5 years) in 1986, from the University of Porto (FEUP), the Aggregate title (Full Habilitation) from the University of Coimbra in 2005, and the D.Sc. from London Metropolitan University in 2013. He is Senior Chartered Engineer by the Portuguese Institution of Engineers with an MBA and Specialist title in Engineering and Industrial Management. He is also Eur Ing by FEANI-Brussels and Fellow (FIET) by IET-London. Currently, he is Professor at the Department of Mechanical Engineering of the University of Aveiro, Portugal. He has more than 30 years of teaching and research experience in manufacturing, materials, mechanical and industrial engineering, with special emphasis in machining and tribology. He has also interest in management, engineering education, and higher education for sustainability. He has guided large numbers of postdoc, Ph.D. and master's students and has coordinated and participated in several financed research projects. He has received several scientific awards. He has worked as evaluator of projects for European Research Council (ERC) and other international research agencies as well as examiner of Ph.D. thesis for many universities in different countries. He is Editor in Chief of several international journals, Guest Editor of journals, books Editor, book Series Editor, and Scientific Advisory for many international journals and conferences. Presently, he is an Editorial Board Member of 30 international journals and acts as reviewer for more than 100 prestigious Web of Science journals. In addition, he has also published as editor (and co-editor) more than 100 books and as author (and co-author) more than 10 books, 80 book chapters, and 400 articles in journals and conferences (more than 250 articles in journals indexed in Web of Science core collection/h-index 50+/7500+ citations, SCOPUS/h-index 56+/10500+ citations, Google Scholar/h-index 71+/16500+).

Part I
Design Thinking

Chapter 1
Introduction to Design Thinking

1.1 Introduction

Presently, there is a greater difference in the requirement of skill set possessed by the workforce than before. The difference has arisen as a result of intense global competition (Shute and Becker 2010). Design thinking has emerged as one of the skill sets in the recent competitive environment. Simon (1996) considers design as one of the central activity in the engineering stream. There are also opinions on the effective engineering programs from the perspective of design. As for instance, McNeil and Beakley (1990) opine that the engineers must be able to meet social needs and requirements through effective design solutions. Design has often been regarded as a natural and universal human activity and is therefore like any other problem-solving skill set. Design process incepts with the realization of the dissatisfaction regarding the current needs of the society, and hence, certain action needs to be undertaken in this regard. As a result, many in the scientific community are designing to provide the solution and hence are acting as designers. However, as highlighted by Braha and Maimon (1997), these scientific personnel are unaware that they have been involved in some kind of design process.

Lack of sufficient scientific base in engineering domain has been pointed by many in the research community (Braha and Maimon 1997). Basic science has been mainly involved in the development of various models that fulfills the requirements of engineering domain. However, the engineering graduates being produced did not possess the required industrial skills and hence were perceived by academia and industrialists as ineligible to be employed in various industries. This lack of skills forced the leaders in engineering domain to identify the key complexities and the resources that can aid in promoting and implementing good design education (Todd and Magleby 2004). As a result, existent engineering courses were revamped by the inclusion of projects sponsored by various industries that provided the students to solve real industrial problems through associated real-world experience and expertise (Bright 1994; Dutson et al. 1997).

K. Kumar et al., *Design Thinking to Digital Thinking*,
Manufacturing and Surface Engineering,
https://doi.org/10.1007/978-3-030-31359-3_1

Business environment has now begun to be attracted by the design thinking approach. The increasing attractiveness can be attributed to the fact that the design of services and products is one of the major components of competitiveness in business environment. Realizing this fact, various companies have committed themselves to adapt to the design thinking approach and hence become design leaders (Dunne and Martin 2006). With the wider acceptability of design thinking process among the design engineers and the managers, there is also a greater demand to involve it in the education domain as it tends to arouse creative thinking in order to solve various problems. Therefore, it is expected that the new creative process in the form of design thinking will positively influence the education curricula. The students will be required to critically analyze and ponder on a real-world problem and then provide solutions logically. The requirement is further underpinned by the ever-growing digital environment, and therefore, educators are required to hone the skills of their students that can meet the twenty-first-century requirement. Newer evolving processes such as teamwork skills, systems thinking, and design thinking (Shute and Torres 2012) are some of the major paths for enhancing the problem-solving skills and hence prepare the students for industrial environment.

The aforementioned evolving skills are based on the theoretical foundations of constructivism, developmental theories, and cognition (Bruner 1990). The major concern is regarding the level of inclination that an individual or the group tends to possess to acquire and adapt to these new evolving skills. Many literature surveys (Dym et al. 2005; Nagai and Noguchi 2003; Fricke 1999) have revealed that design thinking has received a lot of attention from various domains such as design, engineering, and architecture apart from the business realm. There has been array of research articles in the domain of design thinking, and major of the articles in this area has been on the novice design thinkers and expert design thinkers (Ericsson and Smith 1991; Cross and Cross 1998; Goldschmidt and Weil 1998; Do and Gross 2001; Stempfle and Badke-Schaub 2002; Owen 2007). Design thinking has opened new frontiers for strategical development (Liedtka and Kaplan 2019).

The present chapter illuminates the readers with the various characteristics and processes related to the design thinking methodology. Differences between expert and novice design thinkers have also been introduced in this chapter. This chapter also highlights the importance of design thinking process in the enhancement of students' capability to solve real-world problems.

1.2 Design Thinking Process and Its Nature

The source of knowledge generation is majorly accumulated by doing something and as a result thereof by analyzing the results. Hence, it can be concluded that work is the major source of knowledge generation, and therefor, its careful analysis produces knowledge. According to Owen (2007), the working manner of a creative person can be classified into makers or as finders. The creative aspect of a finder is perceived through the discovery feature. The major task of finders lies in understanding the associated phenomena that are not well elaborated and hence finding the explanation. On the other hand, makers synthesize the knowledge they have regarding the newer arrangements, concepts, compositions, patterns, and constructions. The way in which fundamental processes have been used in differentiating between finders and makers, in the similar way professional fields can be differentiated on the basis of other important factors. This will ultimately aid in defining the nature associated with the design thinking process. The content associated with the working of a field is one differentiating factor among the professional fields.

The factors related to the processes and the content can be used in developing a conceptual map which can be defined using two axes (Owen 2007). An analytic/synthetic axis divides the conceptual map into two halves horizontally. Manner of working of professional fields is differentiated as such. The fields on the left side and right side of the axis concern, respectively, finding and discoveries and inventing and making. The conceptual is also divided into two halves by a symbolic/real axis vertically. The fields in the lower half of the map concern the real world and depict the necessary systems for management of physical environment. The fields existing in the upper half map associate the symbolic world and the required tools such as the policies, institutions, and languages that can aid in effective communication and manipulation of the information.

The division of conceptual map by the aforementioned axes therefore results in four quadrants. The first quadrant is referred to as analytic/symbolic and encompasses domains that are analytic to a great extent such as science, i.e., the fields that have processes and content that are more symbolic and less real. Another way of perceiving is that the subject matter is in abstract form when carrying out any analysis. The second quadrant is referred to as synthetic/symbolic and consists of fields that have synthetic processes and symbolic content. The prominent example includes that of law wherein associated contents such as the social relationships and different policies are symbolic and major of its policies are mainly concerned with the development of laws. The third quadrant is known as analytic/real and encompasses fields that have real contents and analytic processes. As for instance, the medical domain falls into this quadrant as it concerns with the real-world problem of human health and hence is focused primarily on the diagnostic processes. The last quadrant is referred to as synthetic/real and involves fields that have real content and synthetic processes. As for instance, design field falls into this quadrant as it is concerned strongly with the real-world subject matter and is highly synthetic in terms of process scale. However, design is also perceived to have symbolic component as it deals with symbolism and communications. Furthermore, design also has an analytic component since to perform synthesis it requires analysis.

Design can also be modeled in terms of two interdependent spaces with different logic and structures (Hatchuel and Weil 2009). These spaces are known as space of knowledge and space of concepts. All well-established knowledge that concerns to the designers falls into space of knowledge. On the other hand, space of concepts relating to an object are the ones that are not an integral part of space of knowledge i.e., neither true nor false in space of knowledge. Gradual portioning of space of concepts marks the beginning of design process and is continued until the set of space of concept becomes space of knowledge; i.e., in other words a set of objects and items is well depicted by true proposition in space of knowledge. Hence, design process can be considered as a reasoning activity that incepts with a concept relating to an object that is unknown partially and is then expanded to a completely newer knowledge or a newer concept or concept and knowledge.

Liu (1996) defines the process of design thinking surrounding the perception of a designer, i.e., the way in which designers perceive and consequently their thinking process. It has been considered as an iterative and interactive process (Do and Gross 2001) wherein designers tend to explore the representations if any associated with the problem-solving concepts and ideas. In the subsequent steps, they try to establish a relationship between the identified ideas and concepts to solve the problem. In the next step of the iterative framework, the view drawn from the above steps that can aid in the subsequent iterative design efforts is analyzed. The design process often incepts with the diagrammatic depiction of a problem. Details are added to this diagrammatic depiction, ultimately translating it into a more complex representation. Reflections, self-critique, and dialogues of designers involved in design thinking process are represented by the design diagrams. The design diagrams therefore serve to represent the purpose and intent of the designer and are considered as a vehicle primarily intended to solve problems (Nagai and Noguchi 2003).

Design process has been considered a generic process (Braha and Reich 2003) wherein the design requirements and the related specifications are modified by the involved design engineers on the basis of the available information. As a result, the generic process aids in removal of discrepancies if any and hence establishes a fit between the problem space.

Suwa et al. (2000) define design process as a situated act which motivates the designers to invent as per the issues and the requirements surrounding the environment in which they tend to operate. A strong bidirectional relationship between the issues and requirements and the discoveries have been revealed by the authors. Unexpected innovations or discoveries are perceived when a designer found something new in an already established solution or concept. Unexpected discoveries are the major driving force behind the inventions pertaining to the requirements and the issues. Inventions and unexpected discoveries tend to support each other in the sense that inventions often result in newer unexpected results. Different activity modes in process of design such as the drawing of sketches and the understanding of issues and requirements related to the design process are realized truly through the outcomes of the above process. The opportunistic nature of the design process is revealed as result of ongoing designer activities related to the design issues and requirements and hence the evolving solutions and concept.

Dörner (1999) opines the existence of several forms of thinking in the designing process. The design process incepts with a certain idea regarding the perception on the design or the product as well as the intended function it needs to serve. The initial idea that is vague transform into a clearer image of the product. Initial vague idea comes to the mind of the designer who already known something about the product. The knowledge on the same can form source of analogies. Sketches and models are the other forms of thinking processes that aid in transforming the vague idea into a more clear form. The characteristics feature of a product is clarified through sketches and models and therefore aiding in the formation of specific line of thought. This specific line of thought ultimately aids in the development process and lays the foundation of design thinking process. "Picture-word cycle" is another form of design thinking wherein the ideas are put in the form of words which helps in more elaboration and clarification on the initially thought out idea. Whatever is the form of cognitive thinking, the design thinker involved in design process should be able to demonstrate their creativity apart from certain specific characteristics. The characteristics associated with a design thinker are discussed in the next section.

1.3 Characteristics Related to a Design Thinker

From the aforementioned discussion, it is clear that it is difficult to decipher the true nature of design thinking process. Another topic of debate that still remains to be elusive is that related to what are the characteristic features of a design thinker. In pursuit of the elusive quest, a number of characteristic features have been identified that play instrumental role in understanding the thought process of a design thinker. The nature of design thinking can also be deciphered through these characteristics. Few of the characteristics of a design thinker as identified by Owen (2007) have been summarized below.

Human and environmental centered thinking: A good design thinker must continuously ponder on as to what and how should a product be created that fulfill the human needs. Together with the human considerations, they should also consider the environmental interests and should design by keeping them as the primary constraints.
Predisposition toward multi-functionality: Designers must have the capability to provide for multiple solutions to a design problem. However, while doing so they should also keep in mind the related specifics.
Ability to use language for effective communication: Design thinkers must have the potential ability to explain verbally the creative process, ultimately resulting in invention. They should be able to explain different relationships that are not visible.
Avoiding the necessity of choice: Design thinkers often search for multiple solutions or alternative solutions before approaching toward final choice or decision making. New configurations are always thought to come up with design solution. The process often results in the combination of best possible choices and avoids final decisions.
Ability to visualize: Designers must be able to visualize the design problem and hence its solution effectively. They should be able to clearly depict their ideas.
Systemic vision: Design thinkers should consider the problem and the associated opportunities as systemic; i.e., the solutions should be arrived at by consideration of different procedures and concepts.

Affinity for teamwork: Designers must possess interpersonal skills that can aid them to communicate effectively across different disciplines and therefore work with people from these different disciplines.

1.4 Design Thinking Processes

Design process has been characterized as a process that is iterative, exploratory, and chaotic (Braha and Reich 2003). The process incepts with a "brief" (Hatchuel and Weil 2009) specification of the product and terminates with an elaborate product description that is gradually refined to obtain a product with detailed specifications. At intermediate stages of design process, specifications and descriptions on a product can be often conflicting in nature. These specifications may change as a result of the unexpected discoveries during the stage of design process. Therefore, design follows iteratively between concepts and problem solutions until a final concrete problem solution is accomplished.

Several cognitive processes are involved at several stages of design process. Following three processes have been identified by Kolodner and Wills (1996): preparation, assimilation, and strategic control. Designers are required to decipher the relevant processes to be focused on as a part of the preparation phase. Specifications of the problem under consideration, the constraints to the identified problem, reinterpretation of ideas, reformulation of problem, and visualization evolve during the preparation phase. The assimilation process on the other hand aids in exploring the meaning to the proposed solution, data, and the observations pertaining to the problem under consideration. The strategic control part is related to making number of decisions such as that related to ideas needing elaboration, the constraints to be relaxed, and setting priorities during the course of design process. They are required to move flexibly and in an opportunistic manner among the associated tasks and subproblems.

Theory pertaining to the tasks of design thinkers has been outlined by Stempfle and Badke-Schaub (2002). The associated theories such as that of creativity and problem-solving and the theories of cognition were looked upon. Selection, exploration, generation, and comparison are the basic elements of a cognitive operation that aids in solving any problem through the design thinking approach. A given problem space is widened through the generation and exploration elements while through the support of selection and comparison elements the problem space is made narrower. Widening of problem space means that the solutions to a real-world problem are evolved or generated and then investigated with due consideration to the defined goal. Iteratively modification of identified solution takes place or addition of new solution is made. The process is continued until an optimal solution is achieved. Narrowing of the problem space on the other hand results in comparison between two ideas of solution and hence the selection of the best idea. The aforementioned elements can be used in understanding the thought process of designer when they work as a team. When working in a team, designers are required to communicate with each other their thinking process.

The model of design thinking was applied by researchers to three teams comprising each four to six participants from mechanical engineering background. The task assigned to the teams concerned the designing of mechanical concept to develop optical device that could be used in projecting the images of the celestial bodies. The teams communicated with a simulated customer thrice in a single working day. The communications made were recorded. On analysis, it was revealed that the engineering teams spent ninety percent of their time in planning of solution and only ten percent of their time in clarifying the goal.

The above findings however differed to that observed by Mc Neill et al. (1998) who analyzed a team comprising of students from electronics background. The analysis revealed that most of time the team spent was on analysis of problem which was followed by the time spent on synthesis and lastly in evaluation. Researchers opine that the conceptual session of design process incepts with analysis of the functional aspects related to the problem under consideration. The designers focus on three important aspects, i.e., function, structure, and behavior as the session progresses. Then, they focus on the cycle involving analysis, synthesis, and evaluation. They finally engage themselves in synthesis of structure and evaluation of behavior of structure toward the end of the design session.

Similar analysis was also carried out by Goldschmidt and Weil (1998) on a team of engineers with industrial background. During the analysis, the nonlinear nature of design thinking process was revealed and also that the participants followed a forward and backward, i.e., breaking down and validating reasoning strategy. From the above studies, the inconsistencies related to the time spent during a process are clear but the findings suggest that there is a learning and comprehending process during the whole process of design thinking approach. This learning process is eventually responsible for the transformation of a novice design thinker to an expert design thinker.

1.5 Comparison between a Novice Design Thinker and Expert Design Thinker

A good design thinker is one who has the potential ability to solve the design problems using different available strategies. The designer must be able to choose the best strategy that fulfills the requirement of the situation under consideration. Apart from the aforementioned characteristic feature, a good designer must be able to clearly define the problem, search actively for the associated information, summarize the identified information into the requirements, prioritize the requirements, and also take care as not to suppress the initially identified or deciphered idea (Fricke 1999).

Protocol studies were conducted to analyze the problem-solving strategy between senior and junior college students in the domain of industrial design. It was observed through the studies that juniors spent most of their time in gathering the related information and therefore were stuck in progressing toward the solution. Seniors, on the other hand, did not face much difficulty in gathering the related information and therefore were able to solve the problem in hand. The seniors were also able to prioritize the activities in the early stage of problem-solving.

Göker (1997) also investigated the differences between expert and novice design thinkers. The differences were investigated for a task involving the construction of machines through computer simulations. It was revealed through the analysis that the experts were experienced in computer simulations and hence relied on their experience rather than inclining toward an abstract reasoning. On the other hand, novices more or less reasoned toward an abstract design concept.

Günther and Ehrlenspiel (1999) also reported similar findings on performing experiment with a total of twenty expert and novice designers involved in the development of mechanical devices. It was revealed that experts were able to clarify their problem in a relatively lesser time in comparison with the novice design thinker. Similarly on conduction of protocol studies on a set of engineering students by Atman et al. (1999), the same findings were reported; that is, the novices spent lot of time in defining and clarifying the problem. This presented a major hurdle in approaching toward a solution, and hence, they were not able to produce high-quality designs. The findings from both the studies corroborated each other.

Qualitative analysis tools were adopted by Seitamaa-Hakkarainen and Hakkarainen (2001) for carrying out investigations on the relationship between technical such as that related to materials and visual designing aspects as for instance patterns, size, color, etc. The investigation was carried out between novice and expert designers who were involved in the weaving designs. Experts in weaving designs considered the technical and visual aspects simultaneously by integrating both the aspects of the design process. They iterated continuously between the visual and technical aspects of the design task and this was therfore one of the associated significant aspects. They carried out a detailed process for searching a design solution by iteratively moving from one design space to another. On the other hand, majority of the tasks performed by the novice designers revolved around the space of composition. The movement to the space of construction was however very rare.

Substantial differences have been reported between the novice and expert architect by Tang and Gero (2001). The findings also were on similar track as that reported by the aforementioned researchers. They adopted retrospective protocol analysis, and the differences were analyzed in relation to four design levels, i.e., physical, perceptual, functional, and conceptual levels. The physical level of design is mainly concerned with the external world and comprises mainly of looking and drawing actions. The visual and spatial features are attended to by the perceptual level of design. The functional level of design concerns to mapping between the abstract concepts and visual-spatial features. The abstract concepts and perceptual actions are processed through the conceptual level of design process. It was revealed through the analysis that experts had better advantage at the perceptual and physical levels in comparison with the novices.

The search strategies adopted by novice and expert design thinkers in solving design problem related to industrial domain were examined by Ho (2001). The research revealed that an expert design thinker decomposed the problem through the decomposing strategies, whereas on the other hand novice design thinker only focused on the problem at the surface level. However, it was also revealed that the novice and expert design thinkers worked through the bottom-up strategies of problem-solving.

Kavakli and Gero (2002) used protocol studies to compare the cognitive actions such as perceptual and looking of an expert and novice architect. Concurrent cognitive actions were investigated by the researchers through the protocol analysis, and significant differences were revealed for the output produced by expert and novice design thinkers. The analysis was divided into segments. Cognitive actions were encompassed within the cognitive segment. Around 2900 actions and 350 segments were revealed in the design protocol of an expert designer, whereas nearly 1000 actions and 125 segments were the part of novice designers. Therefore, each segment was consisted of eight cognitive segments. Protocol of expert design thinker was nearly 2.5 times richer to that of novice's protocol pertaining to the actions and given that equal time was given to both the participants. Also, the session of expert was 2.5 times richer in segments in comparison with that of novice design thinker. Hence, there was more fluency associated with the expert design thinker in relation to the thinking skills in comparison with the novice design thinkers. The cognitive actions of expert design thinkers rose continuously throughout the activity period while that for the novice started at the peak and then followed a declining trend. It was also observed that the experts had a greater control over their cognitive actions in comparison with the novice design thinkers. As a result, experts were able to govern their performance effectively in comparison with the novices.

Differences in behavior between novice designers and experienced designers were studied by Ahmed et al. (2003). Obvious differences were revealed by the study related to the behavioral aspect of freshmen and an expert into design profession. Trial and error methods were used by the novices in the generation of design solution and in implementation of any modifications pertaining to the identified solution. The generated solution was evaluated iteratively through the employability of hit and trial error methods. On the other hand, preliminary evaluations were made by the expert designers before implementing the final design decision and hence their final decision. Integrated design strategies were adopted by the expert designers.

Novice design thinker on the other hand associates itself with the depth-first approach for solving a particular design problem. This means that the identified and explored intermediary solutions are analyzed in depth initially. An expert design thinker considers top-down and breadth-first approaches. Explicit problem-solving strategies are possessed by expert design thinkers and are another differentiating factor between the novice design thinkers.

1.6 Experts and the Design Thinking Process

As observed by Cross (2004), expert designers have a greater potential ability than the novices to solve a complex design problem. Experts capture the important aspect of any complex design problem by inventing issues or the requirements related to design. This is an extra quality of design expert apart from the ability to synthesize the solution that satisfies the given requirements. The manner of approach toward a problem was examined through protocol studies by Lloyd and Scott (1995). It was observed that the approach was related closely to the type of experience and the degree. Inductive reasoning, i.e., use of generative reasoning, was mainly adopted by the expert design engineers while depth-first approach, i.e., deductive reasoning, was the approach adopted by novices. The experts having experience in the related domain of the problem employed certain assumptions/conjectures to approach the solution instead of spending their time on analysis of the problem. This means that the expert designers approach the problem through the aid of previous solution that they had put forth for the problem of similar kind.

The goals and the related constraints are changed by the designers as they approach toward a problem progressively. Experts are flexible enough to select and try different solutions to a problem. However, designers tend to use their concept of principal solution if they unexpectedly come across any shortcomings or difficulties. Initial ideas of design were one of the major influential factors in determining the direction of their problem-solving (Rowe 1987). They avoided to use any new idea and went with their initial idea even when encountered with certain shortcomings/difficulty.

The findings were in line with that from Ullman et al. (1988) who conducted their protocol studies on a group of engineering students from mechanical engineering background. The expert design engineers followed only a single design approach from the very beginning. Furthermore, the researchers observed that the experienced designers tend to stick to their original idea even on the identification of major problems. They did not invent any new ideas and only modified the already proposed idea. Similar conclusions were drawn from the studies conducted on engineering students involved in real-time electronic projects (Ball et al. 1994). There observation revealed that when the researchers identified that they have generated solution that is not satisfactory, they did not approach to propose a newer solution and continued with their original solution. Rather, they tried to develop newer versions of the original solution until and unless satisfactory solutions were generated. Similarly, an analysis of seniors into software design also revealed the fixation behavior of the experienced designers (Guindon 1990).

Conjectures and assumptions are often used by the expert designers in exploration and hence understanding of the problem formulation. It has been revealed from the protocol studies that the designers incept with exploration of the problem and then progressing gradually toward a partial structure (Dorst and Cross 2001). The initial ideas are then generated through the aid of partial structure and formulate a design concept. The partial structure is developed fully through the gradual expansion of partial structure. Hence, their major objective lies in developing a matching solution to a design problem. The problem is comprehensively understood through the development of more than one solution concept. It is clear from the above literature that a designer is required to hone number of competencies to be a more experienced designer.

1.7 Competency Model for Design Thinking

A competency model for design thinking has been proposed by Shute and Torres (2012). The model is a structured arrangement of arrayed set of variables that expands gradually to a more general view as one progress from left to right of the model. The proposed competency model depicts the operationalization of design thinking concept and different activities that would aid in the collection of appropriate data related to the variables in the design space. As for instance, the associated skills with a design variable "Iterate Designs" are creating, tinkering, and testing of the ideas.

The competency model aids the designers in carrying out tasks related to assessment and diagnostics. The activities pertaining to a design problem can be developed on identification of the required skills and the knowledge. These tasks can be developed in line with the variables of the design problem. However, one of the major challenges concerning a design thinker is that whether the identified skills can be learnt. This major challenge could be overcome through sufficient practices in a conducive environment underpinned by appropriate feedback process. Furthermore, the skills of those involved directly can be enhanced through inquiry-based learning, project-based learning, and problem-based learning (Dym et al. 2005).

The different pedagogical approaches can aid in the creation of awareness related to the conceptual framework of a good design process among the novice designers. This will result in enhancing their capability to solve complex design problems. The design for the associated tasks should be done in such a way that novices can learn to generate ideas and initial solutions. Furthermore, they feel motivated to develop their design thinking skills. Multiple opportunities should be identified by the educators in the domain of design thinking and should support their students in developing their design thinking skills. The novices should be made to collaborate with others, experiment with their ideas, and iteratively revise and improve on their ideas.

Students will be able to come up with more innovative ideas and think innovatively if they are allowed a free hand to apply methods and ideas that come to them. Further, they should be allowed to collaborate with expert designers so that they can learn their approaches toward solving a design problem. Design thinking has been considered more than a skill that can be used in only limited contexts.

1.8 Conclusion

The present chapter provides an overview of the concepts related to design thinking. Literatures have revealed that expert designers are focused centrally on designs rather than being focused on the problem. This is one of the characteristic features of design thinking that only comes with experience and knowledge in the domain of design thinking. Experience in the problem domain aids the designers to identify the problem in a shorter span of time, and hence, proposing an initial solution becomes a tad easier. Synthesizing and evaluation of the generated solution is another feature of design thinking process. Designers should be able to assess suitably the condition associated with the given situation and should be flexible enough to adjust themselves in accordance with the set of requirements.

Students can be well prepared to deal with complex situations and hence solve them if they are prepared to think in a way as designers do. However, the educational system focuses on enhancing students' ability to mater themselves in solving pertaining to traditional subjects. The current educational practices should be modified to transform the pedagogical approach to one that considers valuable skills such as design thinking and digital literacy.

References

S. Ahmed, K.M. Wallace, L.T. Blessing, Understanding the differences between how novice and experienced designers approach design tasks. Res. Eng. Des. **14**(1), 1–11 (2003)

C.J. Atman, J.R. Chimka, K.M. Bursic, H.L. Nachtmann, A comparison of freshman and senior engineering design processes. Des. Stud. **20**(2), 131–152 (1999)

L.J. Ball, J.S.B. Evans, I. Dennis, Cognitive processes in engineering design: a longitudinal study. Ergonomics **37**(11), 1753–1786 (1994)

D. Braha, O. Maimon, The design process: properties, paradigms, and structure. IEEE Trans. Syst. Man Cybern. Part A Syst. Hum. **27**(2), 146–166 (1997)

D. Braha, Y. Reich, Topological structures for modeling engineering design processes. Res. Eng. Des. **14**(4), 185–199 (2003)

A. Bright, Teaching and learning in the engineering clinic program at Harvey Mudd College. J. Eng. Educ. **83**(1), 113–116 (1994)

J.S. Bruner, *Acts of Meaning*, vol. 3 (Harvard University Press, 1990)

N. Cross, Expertise in design: an overview. Des. Stud. **25**(5), 427–441 (2004)

N. Cross, A.C. Cross, Expertise in engineering design. Res. Eng. Des. **10**(3), 141–149 (1998)

E.Y.L. Do, M.D. Gross, Thinking with diagrams in architectural design. Artif. Intell. Rev. **15**, 135–149 (2001)

D. Dörner, Approaching design thinking research. Des. Stud. **20**(5), 407–415 (1999)

K. Dorst, N. Cross, Creativity in the design process: co-evolution of problem–solution. Des. Stud. **22**(5), 425–437 (2001)

D. Dunne, R. Martin, Design thinking and how it will change management education: an interview and discussion. Acad. Manag. Learn. Educ. **5**(4), 512–523 (2006)

A.J. Dutson, R.H. Todd, S.P. Magleby, C.D. Sorensen, A review of literature on teaching engineering design through project-oriented capstone courses. J. Eng. Educ. **86**(1), 17–28 (1997)

C.L. Dym, A.M. Agogino, O. Eris, D.D. Frey, L.J. Leifer, Engineering design thinking, teaching, and learning. J. Eng. Educ. **94**(1), 103–120 (2005)

K.A. Ericsson, J. Smith (eds.), *Toward a General Theory of Expertise: Prospects and Limits* (Cambridge University Press, 1991)

G. Fricke, Successful approaches in dealing with differently precise design problems. Des. Stud. **20**(5), 417–429 (1999)

M.H. Göker, The effects of experience during design problem solving. Des. Stud. **18**(4), 405–426 (1997)

G. Goldschmidt, M. Weil, Contents and structure in design reasoning. Des. Issues **14**(3), 85–100 (1998)

R. Guindon, Knowledge exploited by experts during software system design. Int. J. Man Mach. Stud. **33**(3), 279–304 (1990)

J. Günther, K. Ehrlenspiel, Comparing designers from practice and designers with systematic design education. Des. Stud. **20**(5), 439–451 (1999)

A. Hatchuel, B. Weil, CK design theory: an advanced formulation. Res. Eng. Des. **19**(4), 181 (2009)

C.H. Ho, Some phenomena of problem decomposition strategy for design thinking: differences between novices and experts. Des. Stud. **22**(1), 27–45 (2001)

M. Kavakli, J.S. Gero, The structure of concurrent cognitive actions: a case study on novice and expert designers. Des. Stud. **23**(1), 25–40 (2002)

J.L. Kolodner, L.M. Wills, Powers of observation in creative design. Des. Stud. **17**(4), 385–416 (1996)

J. Liedtka, S. Kaplan, How design thinking opens new frontiers for strategy development. Strateg. Leadersh. (2019)

P. Lloyd, P. Scott, Difference in similarity: interpreting the architectural design process. Environ Plann B Plann Des **22**(4), 383–406 (1995)

T. Mc Neill, J.S. Gero, J. Warren, Understanding conceptual electronic design using protocol analysis. Res. Eng. Des. **10**(3), 129–140 (1998)

Y. Nagai, H. Noguchi, An experimental study on the design thinking process started from difficult keywords: modeling the thinking process of creative design. J. Eng. Des. **14**(4), 429–437 (2003)

C. Owen, Design thinking: notes on its nature and use. Des. Res. Q. **2**(1), 16–27 (2007)

P.G. Rowe, *Design Thinking*, vol. 28 (MIT Press, Cambridge, MA, 1987)

P. Seitamaa-Hakkarainen, K. Hakkarainen, Composition and construction in experts' and novices' weaving design. Des. Stud. **22**(1), 47–66 (2001)

V.J. Shute, B.J. Becker, *Innovative Assessment for the 21st Century* (Springer, New York, NY, 2010)

V.J. Shute, R. Torres, Where streams converge: using evidence-centered design to assess Quest to Learn, in *Technology-based Assessments for 21st Century Skills: Theoretical and Practical Implications from Modern Research*, vol. 91124 (2012)

H.A. Simon, *The Sciences of the Artificial* (MIT Press, 1996)

J. Stempfle, P. Badke-Schaub, Thinking in design teams—an analysis of team communication. Des. Stud. **23**(5), 473–496 (2002)

M. Suwa, J. Gero, T. Purcell, Unexpected discoveries and S-invention of design requirements: important vehicles for a design process. Des. Stud. **21**(6), 539–567 (2000)

H.H. Tang, J.S. Gero, *Sketches as Affordances of Meanings in the Design Process*, ed. by J.S. Gero, B. Tversky, T. Purcell (2001), pp. 271–282

R.H. Todd, S.P. Magleby, Evaluation and rewards for faculty involved in engineering design education. Int. J. Eng. Educ. **20**(3), 333–340 (2004)

D.G. Ullman, T.G. Dietterich, L.A. Stauffer, A model of the mechanical design process based on empirical data. Ai Edam **2**(1), 33–52 (1988)

Chapter 2
Design Thinking in Engineering Realm

2.1 Introduction

Design has often been considered as one of the key activities in the engineering realm (Zindel et al. 2012). Academicians opine that the engineering curricula should be developed so as to graduate engineers who can design effectively so that social needs are fulfilled (Sheppard 2003). Although importance has been given to design, it remains escape for the engineering education. The leaders in the engineering design domain have not recognized the true complexities as well as the resources that can underpin the inclusion of design thinking in engineering education. This is one of the major restrictions identified by the design faculties across the institutions to implementation of design education in the engineering curriculum (Todd and Magleby 2004).

Engineering curricula have been designed on the basis of basic science models. Engineering education is thought to the students only when they gain knowledge on basic sciences and mathematics. The first two years in engineering education focuses mainly on basic sciences and have undergone insignificant changes since the 1950s (Dym 2004). This forms the basic foundation of the engineering students to solve technological problems. Therefore, the students that were graduating from engineering education lacked the required industrial skills owing to the shift from practical to theoretical framework (Dutson et al. 1997). As a result, new courses started to come up in the market as, for instance, capstone design course, i.e., the design course in the USA taken up the seniors in their final year. Such courses have also evolved over the years through the inclusion of projects devised by the faculty experts and the industrialists. These industry-sponsored projects motivate students to work on real problems. The industries support the projects through their technical expertise and financial support (Dutson et al. 1997; Bright 1994).

K. Kumar et al., *Design Thinking to Digital Thinking*,
Manufacturing and Surface Engineering,
https://doi.org/10.1007/978-3-030-31359-3_2

In the 1990s, the students felt the need of engineering faculty since the inception of their engineering course. This resulted in the inclusion of design courses for the first-year engineering graduates. The course was referred to as cornerstone design course (Dym 1999). The students were able to expose themselves to learn the activities of an engineer (Dally and Zhang 1993; Froyd and Ohland 2005). They were able to learn the basic elements of design process with the aid of real design projects (Dym 1994a, b, 2004).

Although significant improvements have been observed in the engineering curriculum through the activated role and the presence of design courses, leaders in the field still feel that further improvements are still necessary (Todd and Magleby 2004; McMasters 2004). Proposals have been put forth to adjudge the goals achieved by the design-based curricula such as the Conceive Design Implement Operate initiative from MIT (Fries 2014). The various assessment measures are a result of consistent debate on the effectiveness of design content in the engineering curricula. The subsequent sections in the chapter incept with basic literature on the conceptual engineering, design, and design thinking. Research on project-based learning is presented in the next section, and it delineates its importance as an effective learning tool for students to grasp the basic elements of design. Some of the related pedagogical issues have also been illuminated in the section. The subsequent section provides a discussion on the future research aspects associated with design thinking. The chapter finally terminates by making final recommendations for further study.

2.2 Design Thinking as Divergent–Convergent Questioning

There are umpteen definitions of engineering. According to Sheppard, engineers "scope, generate, evaluate, and realize ideas." Thinking and the way in which engineers embrace the conceptual framework of design process are the major focuses of Sheppard's characterization. These were highlighted through creation, assessment, selection, and finally bringing up the generated ideas. Knowledge related to technical systems alone cannot aid in understanding the thought processes that can result ultimately in synthesis of design. However, studying and careful analysis of the thought processes are often perceived as a means to improve methodologies associated with design framework (Pahl 1997).

Engineering design on the other hand has been defined by Dym et al. (2005) as a "systematic and intelligent process in which designers generate, evaluate, and specify concepts for processes, systems, and devices whose function achieve the very objectives and needs of user while satisfying certain set of constraints." This very definition is answer to questions such as the meaning of design in the milieu of engineering, its complexity and hurdles in teaching the subject, and so on.

Therefore, engineering design can be considered as a thoughtful process that depends on design concepts and ideas that are generated as a result of intelligent and systematic cognition. Creativity on the one hand is important, and even teachable, design process cannot be considered as invention as portrayed in the instance of flashing light bulb. Design problems are solved by designers who keep in mind the clients and the set of users while approaching toward a solution. Process of design is complex and requires intelligent cognition.

Design thinking process can be characterized through many informative approaches. The following characteristics often highlight the skills of a good designer: ability to think and communicate in different languages of design, ability to judiciously handle any associated uncertainty, ability to make decisions, capability to tolerate the associated ambiguity that arises as a result of viewing design as an inquiry and should be able to track the big picture by including systems design and systems thinking.

Problem definition through questioning marks the beginning of any design project. The designers do not really start with defining the objectives in order to achieve the goal until and unless they assure themselves about the requirements of their client. Questions such as that related to safety of the product, economical aspects, and so on form the integral part of any design project.

The major system of educational font in today's world is referred to as epistemological and encompasses systematic questioning. That said, acceptable solutions are reached for a design problem through some known principles. The major question arises as to whether systematic questioning can result in suitable and acceptable solutions. Researchers answer to this question affirmatively as they opine that the manner in which questions are asked and the way in which the solutions are perceived and presented are the major tools and techniques that can assist the designers effectively (Dym 2004). The questioning process takes place at various stages of design process since the basic design model is an iterative process and is therefore comprised of many stages or phases.

Aristotle has put viewpoint regarding the relationship between questioning and knowledge that one has. Aristotle opines that the kinds of questions asked depend on the knowledge possessed by an individual (Kistler 2006). That is, knowledge can be predicted through the questions an individual can ask and the answers that can be provided to the questions asked. Hierarchy in the questioning perception of Aristotle has been identified by Dillon (1984). That is, some questions are there that needs to be asked before others can be answered. As for instance, the questions pertaining to a phenomenon cannot be asked or it will be misleading to answer questions pertaining to phenomenon before its very existence. Therefore, there is a set procedure in accordance with Aristotle that encompasses an inquiry process which is required to be followed to reveal solutions to questions. Taxonomies pertaining to the inquiry process have been also made available to the computational models (Lehnert and Lehnert 1978). This has also affected the relationship between asking questions and the knowledge residing in individual (Graesser and Person 1994). The types of questions asked during the tutoring process by students have also been influenced through the inquiry process (Graesser and Person 1994).

The Aristotle's approach of asking questions has been extended to the engineering curricula that have become the major strength of the curricula. Students are made to understand the approach of Aristotle through the effective engineering curricula. The deep reasoning questions which have been often termed as a special class of questioning have been correlated affirmatively with the student learning process in the realm of science background. This correlation has been effectively measured and evaluated by a unique test score (Eris 2004). Deep and abdicative reasoning questions, if pertains to the reasoning and comprehending about any specific material, then they should precede the existence of phenomenon in accordance with the Aristotelian framework of approaching engineering problems.

The difference between systematic questioning and epistemological process of inquiry has been observed and analyzed by considering the manner in which designers think and question (Eris 2004). Designers asked over 2000 questions pertaining to a series of quasi-controlled laboratory experiments. These questions were asked by three teams of design engineers comprised of thirty-six designers each. Thinking process of designers was revealed to be unique and identifiable through certain questions that were not placed in any of the published taxonomies of questions.

From the above discussion, it is clear that there exists a specific set of answers for a given design question. This is significant of convergent characteristic pertaining to questioning. The designer involved in convergent thinking tries to converge and henceforth approaches toward revelation of facts. Questions arising out due to such thinking process are associated with truth-value. Deep reasoning questions are examples of such questioning.

For any question in a design process, there are numbers of possible answers that are known which can be either true or false. Together with the known answers, there may be certain unknown answers. Therefore, the questions often operate under a premise that is known as being diametrically opposite in their characteristics. The known alternative answers are usually disclosed by the questioner giving rise to unknown possibilities of answers. This is significant of divergent thinking approach wherein the questioner diverges from the facts in order to create feasible answering possibilities. These questions were also termed as generative design questions (Eris 2004). In such a questioning approach, a designer is not at all concerned with the truthfulness of the answers.

Knowledge domain forms the main operational base for convergent questioning while on the other hand concept domain is the operational base for the divergent questions. This calls for instructing the convergent design thinking process as the so-called concept-knowledge theory (Hatchuel and Weil 2003) is proponent of the fact that truth-value may not be associated with the concepts but knowledge does associate with itself the truth-value. Through such a process of questioning, the designers also upgrade to the existing knowledge base (Vincenti 1990).

Combination of generative design approach as well as deep reasoning relates to the performance in obtaining solutions to the design questions. This fact underpins the significance of transformation between knowledge and concept domains. There are also evidencing documents that reveal that the contents pertaining to the questions correlate truly with the design performance (Mabogunje 1997, 1998).

Hence, an effective design thinking process encompasses both the convergent component and the divergent component. The convergent component aids in the development of deep reasoning questions by slowly building on the lower level and systematic convergent questions. The divergent component on the other hand results into creation of generative questions that aids in concept creation for the convergent component to act.

Teaching divergent component is neither recognized clearly in the design thinking process nor does this performs well in the curricula associated with engineering. As for instance, it is not all acceptable to expect a student to answer a question by teaching multiple concepts that do not hold any truth-value. The students should be motivated to work toward creating unique question by engaging themselves in convergent process. Ability of student to converge is being perceived positively and credit is given to their thinking process irrespective of the correctness of the answers.

With all the above discussion and the context, engineering curricula can be characterized as follows: the engineering curricula should effectively convey Aristotle's convergent process of inquiry being characteristic of epistemological framework. Engineering curricula should not only promote the ability to reason about mathematics and sciences but must also provide the ability to student to answer to the real-world problems.

The main question that arises is the suitability to identify divergent thinking as well as design process that can iteratively transform to divergent–convergent process that can be ultimately used in the development of better pedagogical approach being acceptable to both engineering analysis as well as engineering design.

2.3 Design Systems-Related Thinking

Designers have helped in the development of engineering projects characteristic of being large and ambitious, resulting ultimately into complex world (Stevens 2016). The engineering products are now built with increasing number of components and thereby increasing their complexity. This has been made possible through the increased robustness provided by the designers. Gradually with increased robustness, requirements on environment and social impacts are also required to be fulfilled. Therefore, these requirements call for designers to have skills that aid them to cope with the ever-increasing complexity. As a result, various universities are now coming up with specialized design programs and the associated area (Ng 2004). The present section of the chapter aims to review the systems design and skills pertaining to systems thinking that are required to be possessed by engineering students and therefore exhibited by good design engineers.

System dynamics-related thinking: Anticipation of consequences that could emerge as a result of interaction among the different parts of a system is one of the key characteristic features of a good design engineer. This anticipation can aid in effective management of the designing processes and hence the engineering systems. However, it is very difficult to learn these skills and hence is uniquely possessed by good designers. As for instance, students in one of the schools were not able to reason effectively about the dynamics of simple systems such as that of inventory falling and rising and that of filling and draining of water in tubs (Sweeney and Sterman 2000). It

has been revealed that simpler tasks could be handled well through basic knowledge on engineering education; however, the very elementary engineering education was not sufficient for the difficult task. Many methods of teaching have been proposed that can result in enhanced abilities of people to qualitatively reason about the feedback, outflows, and inflows pertaining to the design system. The people directly involved in the design process have been exposed to what is referred to as "beer game" so that they can have experience of the unintended consequences arising out from the system dynamics. However, no major improvements in the performances were observed in certain group of subjects. As a result of such unresolved difficulties, researchers have proposed to develop research agenda that can aid in enhancing people ability to reason about the systems dynamics. This was achieved through better scientific understanding of thinking associated with system and experiences undergone through such developments (Doyle 1997). The development task could be taken as a research subject by students in engineering discipline.
Associated uncertainties and their reasoning: The process of engineering design incepts with incomplete, imprecise information, imperfect models and objectives that are ambiguous. Such uncertainties significantly affect the design of the engineering systems. Underemphasizing on inclusion of statistics and probability in engineering curricula has been highlighted by different researchers (Hazelrigg 1994). The tendency of people to commit serious errors in probabilities, regression analysis, and the related areas have been reflected through numerous studies carried out in the domain of cognitive psychology (Kahneman et al. 1982). Although knowledge on probability and statistics will result in minimized errors, it will have not to aid significantly in the estimation of uncertainties systematically (Winkler 1967). Conceptual understanding of people on statistics was not revealed to improve significantly through statistics concept inventory (Stone et al. 2003). The effectiveness of communication can be enhanced by the interpretation of outcomes in terms of frequencies rather than in terms of probabilities (Gigerenzer and Hoffrage 1995).

Leaders in the domain of engineering education and the associated disciplines have been involved in working out the related difficulties through employability of hands-on teaching methods and through the aid of graphics and simulations (Ramos and Yokomoto 1999; Dym et al. 2006). However, many researchers feel that there is a long mile to go in considering uncertainties as the key elements of design thinking education (Wood 2004). However, they suggest the following possibilities through which the aforementioned key objectives could be achieved: employment of courses for statistics and probabilities in the preliminary education phase, including the associated uncertainties in the design curricula, through consideration of uncertainties in humanities and technical courses and by giving due consideration to the experimentation and other hands-on projects during design activity. However, the above changes may not be sufficient until and unless research and developments; hence, continuous progress is made in the domain of statistical and probabilistic thinking. Utilization of better and modern computational tools that can underpin probabilistic thinking approach. Another path is to influence the research studies into cognitive psychology and identify the human weaknesses associated with probabilistic and statistical reasoning. This can be achieved through understanding and exploitation of human strengths pertaining to activities such as pattern recognition and visual processing.

Estimate development: The system goes beyond the scope of a designer when the number of decision and design variables and hence the associated interaction grows. This is one of the major challenges associated with system design. The designer is not able to grasp all the entire related details simultaneously. Therefore, one of the major ways in which the system can be brought under the control of designer's capability is by limiting the number of associated factors. A good design engineer is usually good in estimating things such as they have the potential ability to efficiently estimate the size of design parameters and ignore those that can be easily neglected without causing any hindrance to the design activities. However, it has been revealed that the engineering students are not good at estimating things. This was reported through a test administered on a set of engineering students wherein they were asked to estimate physical quantities such as the energy stored in a battery (Linder 1999). The answers to the estimates varied greatly among the students. The sole reason for this poor performance was attributed to the weak understanding of the basic concepts associated with the domain of engineering and hence the inability to form analogies that are appropriate. Current engineering curricula focus only on making precise calculations and do not emphasize on skills related to making approximations (Linder 1999). Therefore, inventing and developing approximation skills is another frontier for open research.

Conduction of experiments: Application of mere scientific principles is not enough to design a system. Design of systems also entails the employability of empirical data and possible experimentations. This has called for designers to learn the design of experiments approach so that they have the potential capability to plan the experimentations and hence analyze the possible outcomes. Six Sigma programs are widely adopted by the industries to teach the techniques that can result in efficient planning and analysis of the experiments. In education area, these techniques are included in college and professional education systems. Significant impact is observed on the industries through dissemination and adoption of various techniques associated with the design of experiments. However, the engineers cannot alone learn effectively through statistical approaches of designing experiments. Deleterious effects on learning have been feared with overdependency on statistical measures of optimal design (Box and Liu 1999). Engineers must alternate their learning between inductive and deductive processes; that is, they should enhance their physical understanding on engineering models in order to suitably understand the experimental approach and then employ this understanding to update the existing one and the associated models. This calls for another research scope of how to effectively teach engineers in making coordinated use of experiments and the related engineering models.

Making effective design decisions: Designers take decisions throughout the design process and as such to aid in making effective decisions many methods and frameworks related to design have been developed by researchers in recent years (Hazelrigg 1999; Pugh 1991; Otto 2003; Rowe 1987). The basis on which these frameworks and methods are based is that design has been considered as a rational process and is undertaken to choose the best among the best design alternatives. Questioning pertaining to the mathematical and scientific soundness. It is necessary to develop the mathematics of design in order to realize engineering design methods and approaches as a true process with associated rationality that produces the best possible outcomes. Mathematics of design must be developed keeping in view that it is a design-intensive process, and therefore, the associated theories must be imported from other fields such as decision theory and economics (Hazelrigg 1999). Hazelrigg developed set of axioms for designing and constructed a model through the derived theorems. The developed statistical model took into consideration uncertainties, information, external factors, risk, and preferences, i.e., basically the elements of game theory. The developed model was able to obtain numerous decisions among which the only one would be the best. The axiomatic approach was argued to yield representations that are more accurate and produces results that have higher probability of winning in a situation that is engrossed with intense competition.

A design-centric view was articulated by Dym et al. (2006) and employed a deterministic model that takes into consideration any ambiguity through the aid of optimization theory. They stressed that the associated goals are key characteristics feature of any design process or approach and calls for making decisions and the manner in which they should be achieved. The key fundamental consideration is the exploration of the relationship between design decisions and the performances arising out as a result of design solutions. The optimization process being used to embed goal and objective-seeking directly into the process of exploration.

Application of the design-centric view to choose the best design concept among number of design concepts was demonstrated by Dieter (1991). A decision matrix was constructed for the determination of the worth associated with the different design concepts. Utility theory formed the basis of Dieter's methodology that formulizes in creation of key values to making decisions. The proposed methodology was similar to that of Pugh selection chart years (Hazelrigg 1999; Pugh 1991; Otto 2003; Rowe 1987). Probability theory was also used by Dieter in order to demonstrate the formulation of decision trees to aid in selection of design concept.

However, there has been criticism pertaining to the role of decision making in design process and identification of process of design as an integral part of decision making. As for instance, Arrow's impossibility theorem, one of the decision premises, is not perceived to be appropriate model that accurately describes a design process (Dym et al. 2002). Furthermore, a premise of decision theory highlights that a post evaluation of the outcome cannot aid in assessment of quality of the decision made. However, in reality it is hard to imagine and visualize a designer who does not focus on the outcome of the design concept. Furthermore, the design model with integrated decision-making framework assumes that it is only after the generation of the alternatives pertaining to the design concepts, a designer takes critical decisions. The design alternatives generated can be represented in the forms that can be evaluated using the decision-based design framework. The very framework of design cannot put forth any reasoning as to how the associated concepts and the design alternatives have been generated which is considered to be the most difficult and creative aspect of any design thinking model.

The aforementioned limitations have been acknowledged by some of the decision theorists, and they therefore opine that decision analysis should only be practiced after certain threshold. As for instance, Howard (1988) argues that decision analysis tools and techniques must only be applied once the alternatives have been framed and created. This is done in order to ensure that designers are working on the right design problem. Framing was considered by Howard as the most difficult part of the design analysis process. The process of alternative framing requires an attitude that is considered to be unique of human behavior. Automation of the decision analysis in the design arena is challenged by the framing process.

Design thinking in a team: Design process is taught as a team process with due consideration to number of social and technical dimensions (Dym et al. 2003). Learning has often been considered as a social activity by the constructivist theories of learning. Both the capstone and cornerstone project-based courses aid in enhancing the ability of students to work in teams. These project-based courses also aid in improvising communication skills in students. As a result of these outcomes, colleges and universities are incorporating the social and technical dimensions in their design classes (Quinn 1994; Dym 1994a, b; Sheppard et al. 1997; Tooley and Hall 1999; Ulrich 2003).

However, many researchers argue over the advantages and the disadvantages associated with the practice of alternative design concepts. As for instance, Rittel and Webber (1973) emphasize that in the early stages of a design process, designers must raise consistent questions, and hence, design process should be "inherently argumentative" in its early stages. Jentoft and Chuenpagdee (2009) define design as a social process wherein the teams define the problems and then negotiate on these design alternatives. The designers have certain technical values that can aid in filtering while the design interactions are underway. Therefore, the resulting design concept is a summation of participant's perception. A number of pedagogical exercises were developed as a result of the aforementioned design framework as, for instance, the Delta Design jigsaw exercise that ultimately aided in promotion of multidisciplinary discourse and hence negotiation of the associated constraints.

Bucciarelli's views on negotiation and ambiguity were supported by Minneman (1991), and it was opined that these are inherent part of design framework and underpins in understanding as well as structuring of the activities associated with the design process. It was further argued, the focus of the design group get shifted to the communication skills owing to the consideration of aforementioned views which also stimulates the group to consider design education that should be used as a means to teach designers to work effectively in a team environment.

Researchers have also investigated the role played by gender in design team and also in dissemination of design education (Agogino and Linn 1992; Hansen et al. 1995; Cassell 2002; Linn 2005; Rosser 2004). As for instance, examination was carried out on student's perception pertaining to gender in design process through a freshman course carried out at the University of California at Berkeley (Newman et al. 2004). The course encompassed issues related to product development and gender perception toward these issues. Students were asked to participate in workshops in multidisciplinary design teams. The data collection was done using three techniques: questionnaires, interviews, and assignments related to the design process. The examination revealed that the students firmly believed that a good design approach should be able to serve the potential user base including the ones that are underrepresented with the technology traditionally. An increased level of confidence in technology was shown by the students statistically. Furthermore, an increased level of comfort was shown toward working on design projects.

Six diverse factors were taken into consideration by Carrillo and Leifer (2002) to adjudge the diversity on team performance. These diverse factors included technical discipline, years of experience, ethnicity, gender, distance from campus, and Myers-Briggs type. It was demonstrated through the study that the impact of the diverse factors is dependent on time and that its results underpin the case to maximize the diversity. However, it is really difficult to consider the impact of individual diverse factors statistically.

Psychometric measurements associated with the type of personality are a discussion in the design domain and hence design learning. Myers-Briggs Temperament Indicator can be used to predict the likelihood of the success that a team can achieve and also analyze the behavior of the team members (Barrick et al. 1998; Reilly et al. 2002). Such techniques have been often used in the formation of design teams in engineering classes. Jungian typology was adopted by Wilde (2000) to assist in the formation of engineering design teams comprising of students, and on analysis, it was revealed that a design team comprising of candidates with complementary roles, a neutral manager, plurality of viewpoints and a wild card resulted in successful outcomes in comparison with the formations. Satisfaction and cohesion among the team members were considered to be impacted by the so-called collective efficacy (Lent et al. 2002). It was revealed through the examination that outcome expectations may be limited by the negative feelings associated with collective efficacy, and hence, remedial steps are required to further promote the teamwork.

Engineering design language: Design and the associated design knowledge can be represented through different languages from time to time. The design knowledge is cast in different forms to serve multitude purposes. Engineering students perceive mathematics as one of the design languages as it is used widely in solving wider range of engineering problems. Through the forthcoming discussions, it will become obvious that design requires many languages in addition to mathematics. Design knowledge comprises of knowledge on the procedures related to design process, procedures, design objects, attributes of designed objects and shortcuts. Designers think about the process and procedure of design when they incept to sketch and draw the objects they are thinking to be designed. A complete understanding of the concepts related to the design is required in order to represent a design object and its attributes. The role played by design languages in design process has been discussed both in grammatical and philosophical terms (De Souza 1993; Bucciarelli 2002).

Several representations or languages used in design include numbers, mathematical models, shape grammars, graphical representations, and textual statements. Numbers are used to represent design information in discrete terms such as dimensions of parts. These are also used to represent parameters in design calculations or those associated with the analytical or mathematical model. Certain aspects related to the function of design object are expressed through the mathematical models. The behavioral aspect is in turn derived from some physical laws. Shape grammars encompass certain rules to assist designers in combining simpler shapes into a more complex artifact. Language pertaining to features is employed to aggregate specialized geometrical shapes that have associated specific functions. Verbal and textual statements are used to assist in the articulation of design objects and the associated design constraints. It also aids in effective communication between different members of design team and manufacturing teams.

Role of textual languages and their associated aspects in design process has been studied widely by the researchers. As for instance, the relationship between design creativity and the number of phrases or nouns created by the design teams was established by Mabogunje and Leifer. Nouns and phrases generated by the design teams were extracted from their transcripts and the focus was on the unique noun and phrases. These numbers were adjudged to be proportional to the higher levels of creativity.

Analysis of the computational text was carried out to characterize the performance of design engineers (Hill et al. 2001). This complemented with the different psychometric techniques that rely on surveys and interviews. Established methodology provides the self-managing teams and instructors a non-intrusive means to track the behavior of the teams. Therefore, the methodology aids in dealing with the real-time performances of the teams which otherwise in the absence of methodology would have been just after the formation of team or at the postmortem. Examination of the communication, presentation material, and documentation left by the students into design process aided in plotting and hence revealing the histories over design cycle of a product. A positive correlation was established between the outcomes of design and patterns associated with the semantic coherence. Furthermore, similar results were observed when analyzing semantic coherence between the various design stages. The research carried out justified the changing levels of coherence in storytelling during the design stages and also in the design concepts explored by the different design teams. Better performance was revealed for design teams that challenged the assumptions at various stages of design process and therefore had cyclical semantic coherence. This evidences that the teams that have higher performance levels fluctuated between divergent and convergent patterns of thinking.

Analysis of sketching activities through their suitable understanding is the current research topics in the domain of design education. Sketching is a critical and integral part of the design process and is therefore an open research question being challenged to research community. Sketching activities is another design language or representation that aids in storing of useful design solutions and hence highlights the possibilities and the related conflicts. Sketching also forms the basis on which concepts and design ideas are generated. This therefore also facilitates the problem-solving ability of designers (Do et al. 2000). Hence, quality of the design solution is impacted positively as well as the individual experience of design process is enhanced simultaneously (Schütze et al. 2003). Idea generation process is expedited through the sketching activities as they serve an aid for communication, short-term memory, documentation, and analysis (Ullman et al. 1990).

Much of the studies have been published on the impact of sketches on the individual designer but only few have made efforts to analyze the impact on the design team as a whole. Analysis of sketching activities on design teams was carried out by Song and Agogino (2004). The test size comprised of a total thirteen teams with each team encompassing three to seven members each. Four research questions were addressed in the study: useful metrics that can aid in characterization of the design sketches, evolution of sketching-related activities over time, sketches and the relationship with design spaces explore, and correlation between sketching activities and the outcome of the design teams. Variation in sketching patterns was observed through the sketching behavior over the design stages. Significant correlations were also obtained between the metrics identified and the outcome measures.

Type of sketch was revealed to be one of the factors affecting significantly the failure and success of design process through a detailed case study of engineering design carried out by Yang (2003). A significant correlation was established between the quantity of dimensioned drawings and the outcome associated with the graded designs.

Variety was opined to be one of the measures of explored space of design solution during the process of idea generation (Shah et al. 2003). Grouping of ideas was done on the basis of difference between the two ideas; that is, two ideas could be differentiated on the basis of different physical principles. Examinations were also done as to how each function of design was satisfied by the different sets of generated concepts. Variety measure was then applied to the entire group of similar ideas rather than associating variety measure to the individual idea. Significant correlation between total number of drawings and their varieties with the performance of the students was revealed through the study carried out by Song and Agogino. The study also pointed out that the development of best products is dependent on both the breadth and depth of the design space.

2.4 Project-Based Learning

Learning has been motivated and provided with integration through the aid of design projects. Cornerstone courses integrated with project-based learning has resulted in enhancement of students' motivation and also improved their memory retention in the engineering domain. This can be attributed to the fact that project-based learning tends to include the related contents and the experiences early in the engineering curriculum. Such courses provide for direct contact of first-year students with the engineering faculty.

An analysis of how engineering students learn and develop themselves was carried out by Brereton (1999). The development process entails the shifting of students' focus from engineering theory to interacting with the hardware systems. It was opined that the learning process takes place at a juncture where there is continuous translation between the abstract representations and the different hardware systems. The process is itself suggestive of convergent–divergent thinking in the context of project-based learning.

Both the types of courses, i.e., capstone and cornerstone, have provided for design experiences representing Kolb's model of experience learning (Kolb et al. 2001). The very notion of Kolb on experience learning and the fact that engineering projects can be truly realized only through the efforts and the focus on real projects motivate the leaders in engineering education to replicate the experiences in the classrooms. This is also corroborated by the scientific and research community notion on a dialectic between hardware and software models. Inclusion of project-based learning approach has been an innovation in the domain of design pedagogy.

An emphasis on reforming engineering education was brought to light by the National Science Foundation report in the year 1997 (Dym et al. 2006). The report also emphasized on project-based learning, teamwork, and close interaction with the industry experts. The need for reforms in engineering education was felt by employers who not only stressed upon the needs of engineers but also on them possessing effective communication skills and capability to be effective team members and ardent learners (Katzenbach and Smith 2015). Aalborg University, Denmark, was the first educational institute of higher education in the world that was founded on the premises of project-based learning approaches (Kjærsdam and Enemark 1994; Dym et al. 2006). Project-based education was seemed to be multidisciplinary and as per the premises of Aalborg, the very education system could be divided into two main themes: problem-oriented and design-oriented. These themes are reflective of the divergent–convergent thinking aspects. The design-oriented education on the one hand deals with know-how, whereas the problem-oriented education on the other hand deals with know-why. Design-oriented education provides for construction of practical problems and design solutions through the knowledge derived from various disciplines of engineering whereas the problem-oriented approach aids in providing solutions of theoretical problems through the employability of any type of knowledge, i.e., irrespective of the source from where the knowledge is derived.

The assessment review of Alborg's curricula based on project-oriented learning revealed excellent results comparable to those obtained from typical courses. However, there was lacuna with regard to the better qualification in cooperation. The very attribute of quality education has been the urgent need of employers, but there has been no published report that speaks on the rating of curricula or schools by individual companies. Although many reputed companies such as Boeing and General Motors have key programs related to assessments in place, none have data published that highlight on the outcomes of school-specific curricula.

One of the key issues in the realm of cognitive sciences, i.e., transfer, has been addressed suitably by project-based learning approach. Transfer has been defined as the ability to extend what is being learnt in one context to the other, new contexts

(Bransford et al. 1999). This key issue has been identified as one of the components of engineering competency. Although design studio has been the protagonist for design thinking outside the engineering realm, it was the medical community that motivated engineers to think globally about the project-based learning approaches. The first-year medical students were revealed to have traits of better practitioners in comparison with those that were only taught by lectures (Barrows 1985). In today's world, various professions are converging, i.e., engineering, law, medicine, and business are drifting toward new paradigm of project- and problem-based pedagogic frameworks. Project-based learning encourages and supports teamwork and enhances knowledge retention ability and design thinking among the novice designers. However, there is dire need to expand and extend the obtained results to different scenarios of design. The demonstration of the values associated with the project-based learning in various design scenarios is another open research frontier. This entails participation across disciplines that encompass temporal as well as geometrical boundaries.

Diversity in implementations of project-based learning in first-year engineering curricula has been revealed through the emerging design-based cornerstone courses. These courses have been extended to various universities wherein students have worked in teams throughout their semester on an external project. At the end of their project, the assessments were done on the basis of final presentations encompassing graphical and colorful representations. Theoretical lectures on design methodology, team ethics, and the associated topics were imparted to the students in tandem with the project. Gradually the courses are disseminated through the studio mode that packages to motivate students to work on three different projects, i.e., one provided by the teaching faculty in order to demonstrate the different design tools, reverse engineering project (Sheppard 1992), and a design project for an external client.

A greater similarity has been established between cornerstone courses and capstone courses by researchers. The only difference that arises is the inclination of cornerstone courses to focus much on conceptual methods of design and lesser on the discipline-specific artifact. The so-called cornerstone courses enhance students' interest in the domain of engineering and enhance their retention in engineering to take up courses in upper levels of their curriculum.

An increase in five-year graduation rate was revealed through a pilot curriculum integrating humanities, social sciences, engineering subjects, and physical sciences. The enhanced graduation rate was attributed the mentoring by senior faculties and sense of community developed via such pilot courses (Olds and Miller 2004). The retention rates for students undertaking engineering courses during their first year were reported to be enhanced by as much as sixteen percent (Richardson and Dantzler 2002). The reasons for this enhanced retention rate were attributed to the development of community sense and mentoring by senior faculty members.

Significant role of project-based courses was revealed in attaining the aforementioned results (Tolle et al. 2004). In majority of the courses wherein the curriculum was changed, design courses were not exclusively part of such changes; instead, projects were included generally in the changed curriculum. Impacts of such courses on the aforementioned results were studied for various universities. All the elements of a design course such as experimental testing, experimental analysis, oral communication, project management, written communication, and working in multidisciplinary teams were elemental part of such changed curriculum (Knight et al. 2003). Significant changes in the retention rates were observed where it increased from three percent to fifty-four percent. Although such gains cannot be solely attributed to these design courses, the students have been provided with the potential ability to make informed choices of the disciplines. Students adopting for such courses have been able to gain valuable experiences (Jamieson et al. 2001).

There is no definitive data associated with the effect of design courses on retention rates owing to the multiple viewpoints on the suitable metrics for analysis of retention rates. However, there is enough literature that evidences the potential ability of design course to produce positive changes in the retention rates of engineering students.

There are certain other aspects of design project implementations that have greatly affected the outcomes that changed curriculum can have on engineering students. A lot of publicity has been generated through the inclusion of number of design contests in engineering education (Sadler et al. 2000). Such design contests are often sponsored by different professional bodies such as SAE, ASME, AIAA. These design contests have become integral part of engineering education through their inclusion in the form of extracurricular activities (Sadler et al. 2000).

Another aspect that seeks due consideration is that such changed engineering curriculum is only taught by a relatively small number of engineering faculty members. The faculty members who are generally not used to such pedagogical vision do not find themselves comfortable to adapt to such changes because of the amount of efforts involved and the kinds of activities they are required to perform. Generally, it has been revealed that the design faculties have a more difficult time with the advancements taking place in the engineering education. Furthermore, they find it difficult to accustom themselves in the backdrop of the rewards in the academic environment. Various metrics have revealed that the long-term sustenance of such design courses seems to be problematic irrespective of the positive impact they can have on educational and retention rates.

There are number of constraints under which an engineer in today's era needs to design objects. These constraints may be global, business, and cultural in contexts. Therefore, it is a requirement for educators across the globe to look beyond the limits and disseminate design education with sole objective of developing a global network. A global network of design connects design teams that are geographically dispersed to their clients through the employability of appropriate technological devices.

There have been certain evidences on the results that courses in a globally dispersed network can have. Notable differences between local and global teams have been revealed through a design competition that was associated with that of arc welding. Better documentations were revealed for the products and processes of globally distributed teams in comparison with the local teams (Leifer 1998). Maintenance of an e-mail logbook can underpin student's development of academic identity. Furthermore, their written communication is also effected positively through such e-mail logbooks (Eik-Nes 2002). A local course encompassing a software management platform was transformed to a global one with careful attention to learner-centric pedagogical approach. This centered pedagogical approach has made the course accessible to global students, and as a result, hundreds of industrial clients have been satisfied through the skills possessed by the students globally (Stark and Schmidt 2002).

An anonymous online survey comprising of around twenty questions was held for a global course (Dutta and Weilbut 2002). A positive implication of the global team approach was revealed through the survey. Around eighty percent of the respondents positively replied toward global approach and revealed that they would love to have such global courses again if provided an opportunity. Dissemination of the course through videoconferencing also received positive response from the participating teams. The result was opposed to the common perception that videoconferencing does not aid in enhancing communication skills. Majority of the students claimed that their outlook changed positively toward the things they perceive globally.

2.5 Some Research Questions Associated with Project-Based Design Education

Number of research questions pertaining to project-based design education has opened research frontiers in the domain of teaching and effective implementation of design thinking. New techniques of assessment are required for the new coming skills. As for instance, the teamwork approach that demands for assigning grades to the individual members in a team to adjudge the performance level of teams is in contrast to the existing engineering curricula wherein the students are evaluated as an individual. Therefore, new design approaches are required to assess the new theme of design learning. Some of the objective measures associated with assessing the impact of engineering curricula include enhanced interest of students in engineering, performance of students in upper-division engineering courses, and satisfaction of an employer with the performance of their employees.

Design has been considered as a mechanism for learning and itself a process of learning. Hence, the integrative design and thinking skills can be better understood by the instructor owing to the notion of instrumenting student's performance and design activities. The transition and hence perseverance of classroom as a research laboratory was first established by Angelo and Cross (1993). Formative feedback can be had on students learning through such transitions. The very notion of research laboratory also motivates research on the effect of various variables on the learning process. Research in design as well as the evaluation of the associated learning process has been made possible through various methods and available metrics. Most of these methods have been applied to develop different performance indicators that could be employed in assessing the performance of design teams and hence adjudge the overall environment of a design class. A wide range of research has been conducted by researchers to study the various processes that could be employed to teach design and understand key knowledge and skills associated with design activities and identification of educational practices that can effectively aid in dissemination of design knowledge (Adams and Atman 2000; Adams et al. 2002).

The project-based learning has been practiced widely to achieve the objective of design thinking in engineering classes. However, still there are open research frontiers that can aid in extensive study of project-based approaches and hence its effective implementation. Some of these questions pertain to the identification of problems, projects, technology required for the development of student. The authentication of project-based learning courses to match the real-world industrial environment requires further investigation. The changing proportions with regard to the context of different institutional missions as well as engineering disciplines are yet to be deciphered. The management of design learning teams from multidisciplinary background and their performance needs further investigation. The most important of all the research issues lies in developing a systematic platform to authentically evaluate the performance of design courses. Another research frontier is to identify and apply psychometric measures in understanding and analyzing the behavior of different design teams.

Further research space arises in design pedagogy. As for instance, the establishment of systematic system that can aid in promoting the learning of convergent and divergent thinking process as a part of engineering education. One of the methods can be to use question-centric thinking process that can aid in raising the student's performance and awareness pertaining to the questioning in design process. However, the major challenge lies in integrating the very process of convergent–divergent thinking into the existing curricula of design education.

Designing questions prompts the engineering students to generate concepts by themselves and then to reason about them by asking deep reasoning questions through the generated concepts and thereby offer solutions. Although, if it is possible at all to design such exams, then another challenge lies is that of the possible grading system for such exams. This is challenging owing to the fact that a concept is neither true nor false.

Different case studies have been employed in engineering education to effectively describe the effectiveness with which the analyses have been made useful in understanding the real-world industrial cases by highlighting industry's best practices. In this category, reverse engineering or dissection engineering has become widely popular in today's engineering curricula. Such approaches have resulted in enhanced retention rates of engineering students, improved students' thinking associated with engineered products. However, the major challenge rises in the effective investment that is required to be provided for establishing such laboratories, studio, and shop facilities.

2.6 Conclusion

Present chapter has illuminated the readers with the associated pedagogy models that can aid in effective implementation of design thinking approach in engineering curricula. Different courses have revealed to enhance students' retention, learning, diversity, and satisfaction. Design thinking has been used to address various business challenges (Fuchs and Golenhofen 2019). However, the introduction of such courses in engineering curricula is cost-intensive. The associated costs may be small on a global scale and the human talent that would be lost by not implementing such schemes. Furthermore, the long-term survival of design thinking courses itself possesses questions and is therefore one of the major issues that demand the attention of prospective leaders in the engineering education space. There is a requirement to increase the population of capable faculty members who can effectively teach design and create environment through creation of the required facilities such as design studios. Hence, one of the major recommendations would be to motivate academicians and students to make design pedagogy as their highest priority.

References

R.S. Adams, C.J. Atman, Characterizing engineering student design processes: an illustration of iteration. Age **5**, 1 (2000)

R. Adams, P. Punnakanta, C.J. Atman, C.D. Lewis, Comparing design team self-reports with actual performance: cross-validating assessment instruments. Age **7**, 1 (2002)

A.M. Agogino, M.C. Linn, Retaining female engineering students: will early design experiences help? Viewpoint Editorial, NSF Directions, National Science Foundation **5**(2), 8–9 (1992)

M.R. Barrick, G.L. Stewart, M.J. Neubert, M.K. Mount, Relating member ability and personality to work-team processes and team effectiveness. J. Appl. Psychol. **83**(3), 377 (1998)

H.S. Barrows, *How to Design a Problem-based Curriculum for the Preclinical Years*, vol. 8 (Springer, 1985)

G.E. Box, P.Y. Liu, Statistics as a catalyst to learning by scientific method part I—an example. J. Qual. Technol. **31**(1), 1–15 (1999)

J. Bransford, S. Donovan, J.W. Pellegrino, *How People Learn: Bridging Research and Practice* (National Academies Press, 1999)

M.F. Brereton, The role of hardware in learning engineering fundamentals: an empirical study of engineering design and product analysis activity (1999)

A. Bright, Teaching and learning in the engineering clinic program at Harvey Mudd College. J. Eng. Educ. **83**(1), 113–116 (1994)

L.L. Bucciarelli, Between thought and object in engineering design. Des. Stud. **23**(3), 219–231 (2002)

A.G. Carrillo, L.J. Leifer, Engineering design team performance: quantitative evidence that membership diversity effects are time dependent, Doctoral dissertation, Stanford University, 2002

J. Cassell, *Genderizing HCI. The Handbook of Human–Computer Interaction* (Erlbaum, Mahwah, NJ, 2002), pp. 402–411

K.P. Cross, *Classroom Assessment Techniques. A Handbook for College Teachers* (Jossey-Bass, Incorporated, 1993)

J.W. Dally, G.M. Zhang, A freshman engineering design course. J. Eng. Educ. **82**(2), 83–91 (1993)

C.S. De Souza, The semiotic engineering of user interface languages. Int. J. Man Mach. Stud. **39**(5), 753–773 (1993)
G.E. Dieter, *Engineering Design: A Materials and Processing Approach*, vol. 2 (McGraw-Hill, New York, 1991)
J.T. Dillon, The classification of research questions. Rev. Educ. Res. **54**(3), 327–361 (1984)
E.Y.L. Do, M.D. Gross, B. Neiman, C. Zimring, Intentions in and relations among design drawings. Des. Stud. **21**(5), 483–503 (2000)
J.K. Doyle, The cognitive psychology of systems thinking. Syst. Dyn. Rev. J. Syst. Dyn. Soc. **13**(3), 253–265 (1997)
A.J. Dutson, R.H. Todd, S.P. Magleby, C.D. Sorensen, A review of literature on teaching engineering design through project-oriented capstone courses. J. Eng. Educ. **86**(1), 17–28 (1997)
D. Dutta, V. Weilbut, Team teaching and team learning on a global scale: an insider's account of a successful experiment, in *Proceedings, 2002 Networked Learning Conference*, May 2002
C.L. Dym, *Engineering Design: A Synthesis of Views* (Cambridge University Press, 1994a)
C.L. Dym, Teaching design to freshmen: style and content. J. Eng. Educ. **83**(4), 303–310 (1994b)
C.L. Dym, Learning engineering: design, languages, and experiences. J. Eng. Educ. **88**(2), 145–148 (1999)
C.L. Dym, Design, systems, and engineering education. Int. J. Eng. Educ. **20**(3), 305–312 (2004)
C.L. Dym, W.H. Wood, M.J. Scott, Rank ordering engineering designs: pairwise comparison charts and Borda counts. Res. Eng. Des. **13**(4), 236–242 (2002)
C.L. Dym, J.W. Wesner, L. Winner, Social dimensions of engineering design: observations from Mudd Design Workshop III. J. Eng. Educ. **92**(1), 105–107 (2003)
C.L. Dym, A.M. Agogino, O. Eris, D.D. Frey, L.J. Leifer, Engineering design thinking, teaching, and learning. J. Eng. Educ. **94**(1), 103–120 (2005)
C.L. Dym, A.M. Agogino, O. Eris, D.D. Frey, L.J. Leifer, Engineering design thinking, teaching, and learning. IEEE Eng. Manage. Rev. **34**(1), 65–92 (2006)
N.L. Eik-Nes, Using e-mail logbooks to facilitate scientific publication (2002)
O. Eris, *Effective Inquiry for Innovative Engineering Design*, vol. 10 (Springer Science & Business Media, 2004)
R. Fries, An analysis of first year students' changing perceptions of engineering design and practice. Age **24**, 1 (2014)
J.E. Froyd, M.W. Ohland, Integrated engineering curricula. J. Eng. Educ. **94**(1), 147–164 (2005)
C. Fuchs, F.J. Golenhofen, Creating customer value through design thinking, in *Mastering Disruption and Innovation in Product Management* (Springer, Cham, 2019), pp. 77–102
G. Gigerenzer, U. Hoffrage, How to improve Bayesian reasoning without instruction: frequency formats. Psychol. Rev. **102**(4), 684 (1995)
A.C. Graesser, N.K. Person, Question asking during tutoring. Am. Educ. Res. J. **31**(1), 104–137 (1994)
L.S. Hansen, J. Walker, B. Flom, *Growing Smart: What's Working for Girls in School* (American Association of University Women, 1995)
A. Hatchuel, B. Weil, A new approach of innovative design: an introduction to CK theory, in *DS 31: Proceedings of ICED 03, the 14th International Conference on Engineering Design* (Stockholm, 2003)
G.A. Hazelrigg, *Rethinking the Curriculum: Is Today's Engineering Education Irrelevant, Incomplete, and Incorrect?* (Prism, ASEE, 1994)
G.A. Hazelrigg, An axiomatic framework for engineering design. J. Mech. Des. **121**(3), 342–347 (1999)
A. Hill, S. Song, A. Dong, A. Agogino, Identifying shared understanding in design using document analysis, in *Proceedings of the 13th International Conference on Design Theory and Methodology*, Sept 2001 (American Society of Mechanical Engineers), pp. 9–12
R.A. Howard, Decision analysis: practice and promise. Manage. Sci. **34**(6), 679–695 (1988)
L. Jamieson, E.J. Coyle, W. Oakes, Epics: meeting EC 2000 through service learning, in *2001 Annual Conference*, June 2001, pp. 6–462

S. Jentoft, R. Chuenpagdee, Fisheries and coastal governance as a wicked problem. Mar. Policy **33**(4), 553–560 (2009)
D. Kahneman, S.P. Slovic, P. Slovic, A. Tversky (eds.), *Judgment Under Uncertainty: Heuristics and Biases* (Cambridge University Press, 1982)
J.R. Katzenbach, D.K. Smith, *The Wisdom of Teams: Creating the High-Performance Organization* (Harvard Business Review Press, 2015)
M. Kistler, *Causation and Laws of Nature* (Routledge, 2006)
F. Kjærsdam, S. Enemark, *The Aalborg Experiment Project Innovation in University Education* (Aalborg Universitetsforlag, 1994)
D.W. Knight, L.E. Carlson, J.F. Sullivan, Staying in engineering: impact of a hands-on, team-based, first-year projects course on student retention. Age **8**, 1 (2003)
D.A. Kolb, R.E. Boyatzis, C. Mainemelis, Experiential learning theory: previous research and new directions, in *Perspectives on Thinking, Learning, and Cognitive Styles*, vol. 1, no. 8 (2001), pp. 227–247
W.G. Lehnert, W.G. Lehnert, *The Process of Question Answering: A Computer Simulation of Cognition*, vol. 978 (Lawrence Erlbaum, Hillsdale, NJ, 1978)
L. Leifer, Design-team performance: metrics and the impact of technology, in *Evaluating Corporate Training: Models and Issues* (Springer, Dordrecht, 1998), pp. 297–319
R.W. Lent, L. Schmidt, J. Schmidt, G. Pertmer, Exploration of collective efficacy beliefs in student project teams: implications for student and team outcomes. Age **7**, 1 (2002)
B.M. Linder, Understanding estimation and its relation to engineering education, Doctoral dissertation, Massachusetts Institute of Technology, 1999
M.C. Linn, *Technology and Gender Equity: What Works* (Women in Science and Technology, American Psychological Association, New York, NY, 2005)
A. Mabogunje, Noun phrases as surrogates for measuring early phases of the mechanical design process, in *Proceedings of the 9th International Conference on Design Theory and Methodology* (1997)
A.O. Mabogunje, *Measuring Conceptual Design Process Performance in Mechanical Engineering: A Question Based Approach* (1998)
J.H. McMasters, Influencing engineering education: one (acrospace) industry perspective. Int. J. Eng. Educ. **20**(3), 353–371 (2004)
S.L. Minneman, The social construction of a technical reality: empirical studies of group engineering design practice, Doctoral dissertation, Stanford University, 1991
C.C. Newman, A.M. Agogino, M. Bauer, J. Mankoff, Perceptions of the design process: an examination of gendered aspects of new product development (2004)
C. Ng, Findings from a web based survey of degree programs in engineering systems, in *Engineering Systems Symposium* (2004)
B.M. Olds, R.L. Miller, The effect of a first-year integrated engineering curriculum on graduation rates and student satisfaction: a longitudinal study. J. Eng. Educ. **93**(1), 23–35 (2004)
K.N. Otto, *Product Design: Techniques in Reverse Engineering and New Product Development* (Tsinghua University Press Ltd, 2003)
G. Pahl, How and why collaboration with cognitive psychologists began, in *Designers: The Key to Successful Product Development* (1997)
S. Pugh, *Total Design: Integrated Methods for Successful Product Engineering* (1991)
R.G. Quinn, The fundamentals of engineering: the art of engineering. J. Eng. Educ. **83**(2), 120–123 (1994)
J. Ramos, C. Yokomoto, Making probabilistic methods real, relevant, and interesting using MATLAB, in *FIE'99 Frontiers in Education. 29th Annual Frontiers in Education Conference. Designing the Future of Science and Engineering Education. Conference Proceedings*, IEEE Cat. No. 99CH37011, vol. 3, Nov (IEEE, 1999), pp. 13B4–2
R.R. Reilly, G.S. Lynn, Z.H. Aronson, The role of personality in new product development team performance. J. Eng. Technol. Manage. **19**(1), 39–58 (2002)

J. Richardson, J. Dantzler, Effect of a freshman engineering program on retention and academic performance, in *32nd Annual Frontiers in Education*, vol. 3, Nov 2002 (IEEE)
H.W.J. Rittel, M.M. Webber, Dilemmas in a general theory of planning. Policy Sci. **4**, 155 (1973)
S.V. Rosser, *The Science Glass Ceiling: Academic Women Scientist and the Struggle to Succeed* (Routledge, 2004)
P.G. Rowe, *Design Thinking* (MIT Press, Cambridge, MA, 1987), p. 28
P.M. Sadler, H.P. Coyle, M. Schwartz, Engineering competitions in the middle school classroom: key elements in developing effective design challenges. J. Learn. Sci. **9**(3), 299–327 (2000)
M. Schütze, P. Sachse, A. Römer, Support value of sketching in the design process. Res. Eng. Des. **14**(2), 89–97 (2003)
J.J. Shah, S.M. Smith, N. Vargas-Hernandez, Metrics for measuring ideation effectiveness. Des. Stud. **24**(2), 111–134 (2003)
S.D. Sheppard, Mechanical dissection: an experience in how things work, in *Proceedings of the Engineering Education: Curriculum Innovation & Integration* (1992), pp. 6–10
S.D. Sheppard, A description of engineering: an essential backdrop for interpreting engineering education, in *Proceedings (CD), Mudd Design Workshop IV* (Harvey Mudd College, 2003)
S. Sheppard, R. Jenison, A. Agogino, M. Brereton, L. Bocciarelli, J. Dally, R. Faste, Examples of freshman design education. Int. J. Eng. Educ. **13**(4), 248–261 (1997)
S. Song, A.M. Agogino, Insights on designers' sketching activities in new product design teams, in *Proceedings of the ASME Design Theory and Methods Conference*, Jan 2004, pp. 351–360
C.M. Stark, K.J. Schmidt, Creating e-Learning: a comparison of two development programs, in *Proceedings, e-Technologies in Engineering Education: Learning Outcomes Providing Future Possibilities* (2002)
R. Stevens, *Engineering Mega-Systems: The Challenge of Systems Engineering in the Information Age* (Auerbach Publications, 2016)
A. Stone, K. Allen, T.R. Rhoads, T.J. Murphy, R.L. Shehab, C. Saha, The statistics concept inventory: a pilot study, in *33rd Annual Frontiers in Education, 2003. FIE 2003*, vol. 1, Nov (IEEE, 2003), pp. T3D-1
L.B. Sweeney, J.D. Sterman, Bathtub dynamics: initial results of a systems thinking inventory. Syst. Dyn. Rev. **16**(4), 249–286 (2000)
R.H. Todd, S.P. Magleby, Evaluation and rewards for faculty involved in engineering design education. Int. J. Eng. Educ. **20**(3), 333–340 (2004)
S.W. Tolle, V.P. Tilden, L.L. Drach, E.K. Fromme, N.A. Perrin, K. Hedberg, Characteristics and proportion of dying Oregonians who personally consider physician-assisted suicide. J. Clin. Ethics **15**, 111–118 (2004)
M.S. Tooley, K. Hall, Using a capstone design course to ensure ABET 2000 program outcomes, in *1999 Annual Conference*, June 1999, pp. 4–573
D.G. Ullman, S. Wood, D. Craig, The importance of drawing in the mechanical design process. Comput. Graph. **14**(2), 263–274 (1990)
K.T. Ulrich, *Product Design and Development* (Tata McGraw-Hill Education, 2003)
W.G. Vincenti, *What Engineers Know and How They Know It*, vol. 141 (Johns Hopkins University Press, Baltimore, 1990)
D. Wilde, Design team formation using Jungian typology, in *Proceedings National Collegiate Inventors and Innovators Alliance* (2000)
R.L. Winkler, The assessment of prior distributions in Bayesian analysis. J. Am. Stat. Assoc. **62**(319), 776–800 (1967)
W.H. Wood, Decision-based design: a vehicle for curriculum integration. Int. J. Eng. Educ. **20**(3), 433–439 (2004)

M.C. Yang, Concept generation and sketching: correlations with design outcome, in *Proceedings of DETC-DTM: ASME Design Engineering Technical Conferences—Design Theories and Methodologies*, Sept 2003, pp. 2–6

M.L. Zindel, J. Mello da Silva, J.C.F. Souza, S.B.S. Monteiro, E.C. Oliveira, A new approach in engineering education: the design-centric curriculum at the University of Brasília-Brazil. Int. J. Basic Appl. Sci. **12**(5), 97–102 (2012)

Chapter 3
Methods and Tools of Design Thinking

3.1 Introduction

Design thinking has been one of the key driving forces for innovation and is now widely accepted across many disciplines such as management (Seidel and Fixson 2013). Both the practitioner and scholarly literatures have become inclined to this newer framework of design approach, i.e., design thinking. As already stated, design thinking constitutes of spaces that are overlapping in nature. These spaces are viability space, desirability space, and feasibility space. Business perspective is reflected in the viability space whereas the user's space is reflected in desirability. The third space of design thinking, i.e., feasibility comprises technological perspective. Innovation is enhanced and expedited when the three perspectives have been achieved. Design thinking has the potential ability to solve complex problems referred to as wicked problems (Buchanan 1992). This potential ability has allowed it to be used in business environment and has been considered as one of the possible pathways for innovation. Design thinking process has been aided by various tools and methods of design and therefore supports innovation for design teams comprising of both the designers and non-designers. From the perspective of a designer, the design thinking process encompasses concept generation, attributes of creative process, and methods such as abductive reasoning and rapid prototyping (Kolko 2012). From the business perspective, one of the important components of design thinking methodology is establishment of deep understanding within the members of a design team (Liedtka and Ogilvie 2011).

Business growth is fostered and maintained by business innovations that provide business with competitive advantage in the market. As a result, the businesses tend to not only meet the requirements of their consumer base but have also the potential ability to underpin customization of products. One of the issues that attract the business communities is that of understanding the methods and tools that can

K. Kumar et al., *Design Thinking to Digital Thinking*,
Manufacturing and Surface Engineering,
https://doi.org/10.1007/978-3-030-31359-3_3

support innovation in a team environment. However, there is lacuna in the literature space that highlights the specific guidelines for fostering innovations through various methods and tools of design thinking that can be employed by team encompassing non-designers. The present chapter therefore provides a rich insight into the tools and methods that can be used for design thinking ultimately resulting in the concept of generation and hence creation of innovative solutions. Toward the end of the chapter, certain guidelines associated with the utilization of design thinking methods and tools have been presented.

3.2 Design Thinking and Innovation

Owing to the increasing competition across the global market and rapidly changing demands of market, organizations are required to innovate. Such business challenges have been supported by the design thinking approach that has been attracted both the designers as well as non-designers (Seidel and Fixson 2013). Design thinking has been considered as a successful method for idea incubation during the first phase of innovation (Meinel and Leifer 2012). Number of literatures have been published with due consideration to relationship between design thinking and innovation and the factors that affect the growth of innovation (Beckman and Barry 2007; Gielnik et al. 2014; Seidel and Fixson 2013). Innovation has been considered as a process that is contributed by several factors possessing complimentary capabilities (Harhoff et al. 2003). Innovation has also been defined as a multistage process whereby new products and processes are produced as a result of transformations of various ideas (Baregheh et al. 2009).

Design thinking can also be visualized as the employability of available methods and tools by teams comprising of personnel from multidisciplinary backgrounds to address innovation challenges (Seidel and Fixson 2013). Adoption of design thinking tools and methods was investigated by Seidel and Fixson (2013) for multidisciplinary teams comprising of novices. The study revealed that for wider acceptability of design thinking approach, methods and tools need to be employed by novices, the effect of adoption not known. When the novices in multidisciplinary teams are acquainted with the combinatorial effects of various methods, limits for the session associated with brainstorming and transformation to practices that are less reflective from the more reflective ones, the novices are more likely to successfully apply the design thinking tools. This results in enhanced overall performance of the design team. Therefore, organizations tend to employ multidisciplinary teams with novices for their design projects (West et al. 2003). Process of innovation and its management forms one of the key strategic issues of any organization that largely depends on the formation of multidisciplinary teams. The quality of decision making as well as the ability to solve problem is enhanced altogether through the adoption of multiple design perspectives (West et al. 2003). The relationships between innovation, leadership, and team processes were examined by West et al. (2003) in the context of health care. Different models of brainstorming imply that creativity of a group is influenced greatly from the multidisciplinarity proposition as creative and novel ideas are generated as a result of brainstorming sessions. A higher level of cognitive simulation is also elicited through the contextual framework of different formed groups (Fay et al. 2006). Range of skills with broader outlook, abilities, and knowledge available to collaborative task force is also significant of higher degree of multidisciplinarity.

Design thinking process encompasses divergent and convergent thinking framework that has relevant effect on the team collaboration (Brown and Wyatt 2010). A design team is initially required to broaden their base of thinking, i.e., making it divergent which allows for multiple inputs for the associated problems. The process of divergent thinking process is considered to be a creative process that aids in clearly defining the real problem (Brown and Wyatt 2010; Lapierre and Giroux 2003). A better insight into entrepreneurial aspect as well as any possible lack in experience can be had with this creative phase of design thinking through creation of new tasks and relevant information being searched (Lapierre and Giroux 2003). Process of creativity and innovation entails creation and application of new knowledge and thereby supporting divergent thinking as one of the key attributes for innovation (Lapierre and Giroux 2003). The composition of the team formed significantly influences the process of design at this stage also. The aforementioned discussion therefore reflects the importance of multidisciplinarity in divergent thinking wherein this results in stimulation of creative inputs. Implementation of identified ideas into action through adoption of convergent process of thinking and then employ the innovative solution is the last phase of innovation process.

A study into a workshop using service design tools to incubate ideas, develop prototype, and assess the business ideas was undertaken by Garcia et al. (2013). The study described how the ideas were translated into useful business opportunities. It was argued that the design thinking approach underpinned by aforementioned design toolkit could support and aid in the development of new entrepreneurial ventures. Certain strategies aiding in encouragement of innovation through education and organizational changes pertaining to workspaces were opined by Beckman and Barry (2007). Two phases have been argued for a design process, i.e., analytical and synthetic phase. The analytical phase encompasses to find and discover ideas whereas the synthetic phase results in innovation and creation. Combination of these theories was proposed for creation of innovation through ethnographic and observational research. The combinatorial approach was also argued to support in creation of framework that could aid in understanding the data, analysis of new requirements of customer, and development of new products or solution that meets the needs of customers.

Although the relevance of design thinking process has been acknowledged in generation and supporting innovation, the manner in which the design thinking tools and methods foster the process of innovation needs attention from research and business communities.

3.3 Tools and Methods for Design Thinking

There are a large number of tools and methods available to facilitate the innovative process of design thinking. More than 160 methods and tools associated with the design thinking approach were identified through various sources, i.e., both industrial and academia (Alves and Nunes 2013). Twenty-five selected service design tools were employed for the formation of guidelines that aided novices to effectively work in team environment coherently. This was also argued to be supportive framework for practitioners.

The various identified methods were clustered into different dimensions using a four-quadrant chart. The different dimensions considered were activities in design process, representations employed, the intended audience, and motivation to employ the same. Since majority of these methods are employed for gaining a proper understanding to the design problem, selection of right method is extremely important in the first phase of design process.

Design thinking process can be approached through five different stages. Empathizing reflects the user interaction for which the design solution is meant. The problem identified through user interaction aids in second stage of design thinking process, i.e., defining the problem. Once the problem has been defined, ideation stage kicks in. This stage involves various sessions of brainstorming and hence generating feasible solutions. The next stage is that of prototyping phase that entails the development of number of prototypes. The process then terminates with the testing stage. Design thinking can be perceived to be creation of meaning or making sense of things when out looked in design context. Multidisciplinary teams can have effective decision making and communication through selection of right design tool. The tools can be either software or physical such as pen, whiteboard, and paper. Design tools can also aid the design teams to suitably adopt to new design perspectives or in visualization of complexity associated with a design system. Also, convergent or divergent view of thinking can be reflected on the basis of design stage.

The subsequent discussion is on certain methods of design thinking (Chasanidou et al. 2015). The corresponding Web-based software tools that are used to implement these methods have also been discussed. The criteria of selecting the tools are their ability to enhance communication skills and visualization technique within the members of a multidisciplinary team. Their usage simplicity is another aspect that attracts their selection.

Persona: needs and desires of users can be identified through the aid of persona method. Persona represents a user's representation that intends to simplify the communication skills as well as decision making pertaining to a design project. This is accomplished by the selection of project rules matching the propositions and concepts of the design project (Junior and Filgueiras 2005). Character intent is also represented by persona and therefore is significant of the attitude of design teams and clients to engage themselves in a design process efficiently. The first stage of design thinking process, i.e., empathizing, can entail the usage of persona. This can be implemented through Web-based software tool such as Smaply (www.smaply.

com). The very Web-based tool provides for several options that describe personas and options pertaining to collaborations and visualization.

Customer Journey Map: service blueprinting is the origin for customer journey map (Shostack 1984). A customer journey map describes the collection of touch points throughout the service cycle, i.e., from its inception to its delivery. A potential point of communication between service provider and the customer is referred to as a touch point. Customer journey map helps in identification of problems associated with services and hence probable chances for innovation in service area (Shostack 1984; Krippendorff 2006). This tool of design thinking can be employed during the empathy stage. A Web-based system referred to as Touchpoint Dashboard (www.touchpointdashboard.com) can be utilized for visualization of service user experience. Visual notations are used for unification of team by conversion of information into a customer journey being characteristic of intuitive and data-rich content.

Business Model Innovation: this method aids in exploration of business opportunities and the associated challenges. Business model and the related managerial, operational, and economical decisions are handled by Business Model Canvas (Osterwalder and Pigneur 2010). Business Model Canvas aids in visual representation of business idea, service, or product. Business perspective of a design thinker is usually reflected in the Business Model Canvas and can be used effectively during the ideation phase of the design thinking process. Strategyzer is one of the Web-based tools that are aided in the implementation of Business Model Canvas. Nine building blocks associated with Business Model Canvas are included in the Web-based tool. These blocks are represented through Post-it notes. Strategyzer encompasses economic analysis, engaging interface that underpins effective communication between the users.

Service Blueprint: the roles and processes of stakeholders are represented through simple steps and service delivery flows. The visualization is made simpler through the employability of service blueprint introduced by Shostack (1984). Different actions among the stakeholders comprising of both the customers and service providers are encompassed in service delivery flows. The technical perspectives of design thinker are usually represented through the service blueprint and are also considered to be one of the process-oriented methods for business. Service blueprint is often used during the problem-defining phase wherein it depicts the actions that can be seen directly by the customers, i.e., in-front tasks and the actions that cannot be seen by the customers, i.e., actions in the back office. Creately is one of the Web-based tools that support the service blueprint process.

Rapid Prototyping: one of the quicker ways of enabling visual formations and hence the experimental implementation of the generated concepts is the employability of rapid prototyping process (Liedtka and Ogilvie 2011). It aids in determination of technological feasibility of the generated solutions. Rapid prototyping methods can be used to create prototypes and test the created prototypes. The conversation and feedback associated with product and services is facilitated through effective communication in multidisciplinary teams through the employability of rapid prototyping methods. Rapid prototyping methods provide robustness to the prototyping phase of the design thinking process and therefore reflect more than the technical perspective

of design thinker. Axure is one of the most suitable examples of rapid prototyping software tool. Such software tool provides for prototyping, wire framing and the other necessary interfaces. It has a graphical user interface that aids in creation of different Web sites and applications carrying tutorials for the novices. Design thinkers are motivated to generate quicker ideas and obtain the feedback to improve the design solution in the least possible time.
Stakeholder Map: the groups involved in a project pertaining to a product or service are visually represented through stakeholder maps. Stakeholders comprise companies, users, customers, and other stakeholders (Blomkvist et al. 2010). Business perspective of a design thinker is reflected through the stakeholder map. The connections among the various stakeholders representing their actions can be mapped and analyzed through the utilization of stakeholder maps. Identification of key stakeholders and their relationships has been argued to be important part of problem-defining phase. Stakeholder Circle is one of the examples of software tool that underpins stakeholder maps. The software tool has facilitated to put stakeholders on the management radar and hence their dynamic assessment. The dynamic assessment is key to decipher the changes in the relationships and hence the projects.

3.4 Applicability of Design Thinking Tools Supporting Innovation

Workshops are one of the best ways to investigate the applicability and usefulness of different design tools for innovation. As for instance, Chasanidou et al. (2015) used two different setups for gaining a rich insight into the relationship between innovation and the employability of different design thinking tools. Three Web-based design thinking tools, i.e., personas, customer journey map, and stakeholder maps, were selected for carrying out the investigation. Out of the two workshops investigated, one was hosted by an academic library and the other by university facilities.

The workshop hosted by academic library encompassed six participants, and the workshop was conducted for a duration of two hours. The participants had already participated in design thinking-associated workshops and seminars (Culén and Gasparini 2014). Although the participants were familiar with customer journey map and possessed excellent computer skills, they were not familiar with Smaply. Therefore, the participants were given two tasks, i.e., to transfer a product or service from a workshop held previously and hence develop a new one. The customer journey map aided the participants to envision how a user, i.e., a university researcher, borrows and gains access to download e-books. The six participants were divided into three groups. One author, i.e., the authors carrying out investigations, was also added to each group and was asked to play assistive roles during the processes. Laptops were provided to each group and thereby facilitating effective communication among the groups. The groups were made familiar with the Smaply software tool and then asked to perform the three tasks, i.e., personas, customer journey map, and stakeholder maps. Small breaks were provided to the groups between the tasks.

The second workshop that was conducted by university facilities had seven participants and was also of two-hour duration. The participants involved had utilized design thinking tools and methods earlier but had no prior experience of workshops. In this workshop also, the participants did not had any experience with the software platform of Smaply, but they possessed good communication skills. The major aim of the conducted workshop was to ask the participants to employ the unique software in the form of Smaply for their semester projects. Three groups were framed, and the participants involved were informed about the processes. The participants were then asked to work on the aforementioned design thinking tasks.

On observing the results from the two workshops, positive outcomes were revealed. Active participation of the involved participants was reported during the entire two-hour session of the workshop. The outcomes were reported in terms of the type of thinking of participant, i.e., convergent, divergent, or convergent/divergent, collaboration, i.e., method or tool driven and the multidisciplinary nature of the groups. Technical constraint associated with a tool was the main observing factor in the first workshop. The participants had ample amount of time to acquaint themselves with the user interface of software platform, i.e., Smaply. Creation of personas was revealed to be much easier task in comparison with the other two tasks, i.e., stakeholder map and customer journey map. Technical issues were faced by some of the participants during the performance of customer journey map task. Lack of richness in the Smaply software was reported by the participants besides the technical issues such as space limitations, labels of buttons.

Another topic of observation from the first workshop was related to the overcoming of the constraints associated with the tool. The groups involved in the workshops drifted away from the real task although they worked intensively. As for instance, one of the groups landed with a new solution in addition to the task in hand, i.e., redesigning of the product or the service. The other groups on the other hand were not able to provide newer solutions. A more cooperative pattern was followed by one of the groups in comparison with the other and they shared their ideas intensively in order to fulfill the requirements of the assigned task and hence complete it while, on the other hand, the other groups remained focus on their ability to solve the assigned problem associated with the user interface of Smaply.

Different outcomes were reported from the second workshop. Extensive cooperation during the session was the first main observation. Groups involved were engaged extensively in long discussions about the tasks and the associated aspects. They used their previous experiences in the domain of design thinking methods to frame and conceptualize the tasks. Participants in the group divided their tasks, i.e., one of the members in the group interacted with the software interface while the others engaged themselves in framing the tasks, i.e., the way in which the tasks need to be performed. After accomplishing the first task, i.e., personas, the discussions were open, covering the overall picture of the entire project. The discussions became more animated which concerned the project-related problems. Although technical issues associated with Smaply were faced by the teams, they were secondary. As for instance, a group was not able to delete a stakeholder after creating the stakeholder. The discussions were not associated with the technical issues but limited only to the conceptual level. Different groups involved in the task landed up with newer solutions and outlook on the project.

3.5 Conclusion

Employability of design tools aids in generation and incubation of ideas and therefore creating innovative solutions. Various design thinking tools must be handled judiciously by both the designers and the non-designers. The teams with members from multidisciplinary backgrounds and competency levels succeeded in the application of design thinking methods. The companies and organizations involved can have a different perspective to the development of their product and services with the possession of sound design thinking approach. Innovation can be sparked with methods that are human and business oriented, as for instance the stakeholder maps. Effectiveness of project can be enhanced through the inclusion of both the convergent and the divergent design thinking methods and tools. However, the participants in a team must be acquainted with different design thinking tools and methods to ensure the effective implementation of the design thinking approach.

References

R. Alves, N.J. Nunes, Towards a taxonomy of service design methods and tools, in *International Conference on Exploring Services Science*, Feb 2013 (Springer, Berlin, Heidelberg), pp. 215–229

A. Baregheh, J. Rowley, S. Sambrook, Towards a multidisciplinary definition of innovation. Manag. Decis. **47**(8), 1323–1339 (2009)

S.L. Beckman, M. Barry, Innovation as a learning process: embedding design thinking. Calif. Manag. Rev. **50**(1), 25–56 (2007)

J. Blomkvist, S. Holmlid, F. Segelström, *Service Design Research: Yesterday, Today and Tomorrow* (2010)

T. Brown, J. Wyatt, Design thinking for social innovation. Dev. Outreach **12**(1), 29–43 (2010)

R. Buchanan, Wicked problems in design thinking. Des. Issues **8**(2), 5–21 (1992)

D. Chasanidou, A.A. Gasparini, E. Lee, Design thinking methods and tools for innovation, in *International Conference of Design, User Experience, and Usability*, Aug 2015 (Springer, Cham), pp. 12–23

A.L. Culén, A. Gasparini, Find a book! Unpacking customer journeys at academic library, in *The Seventh International Conference on Advances in Computer-Human Interactions*, Feb 2014, pp. 89–95

D. Fay, C. Borrill, Z. Amir, R. Haward, M.A. West, Getting the most out of multidisciplinary teams: a multi-sample study of team innovation in health care. J. Occup. Organ. Psychol. **79**(4), 553–567 (2006)

L.M. García, A. Deserti, C. Teixeira, Service design tools as frameworks in the generation of business ideas an action research case study, in *2013 IEEE Tsinghua International Design Management Symposium*, Dec 2013 (IEEE), pp. 338–344

M.M. Gielnik, A.C. Krämer, B. Kappel, M. Frese, Antecedents of business opportunity identification and innovation: investigating the interplay of information processing and information acquisition. Appl. Psychol. **63**(2), 344–381 (2014)

D. Harhoff, J. Henkel, E. Von Hippel, Profiting from voluntary information spillovers: how users benefit by freely revealing their innovations. Res. Policy **32**(10), 1753–1769 (2003)

P.T.A. Junior, L.V.L. Filgueiras, User modeling with personas, in *Proceedings of the 2005 Latin American Conference on Human-Computer Interaction*, Oct 2005 (ACM), pp. 277–282

J. Kolko, *Wicked Problems: Problems Worth Solving* (Ac4d, 2012)

K. Krippendorff, The semantic turn: a new foundation for design. ARTIFACT-ROUTLEDGE **1**(11), 51 (2006)

J. Lapierre, V.P. Giroux, Creativity and work environment in a high-tech context. Creat. Innov. Manag. **12**(1), 11–23 (2003)

J. Liedtka, T. Ogilvie, *Designing for Growth: A Design Thinking Tool Kit for Managers* (Columbia University Press, 2011)

C. Meinel, L. Leifer, Design thinking research, in *Design Thinking Research* (Springer, Berlin, Heidelberg, 2012), pp. 1–11

A. Osterwalder, Y. Pigneur, *Business Model Generation: A Handbook for Visionaries, Game Changers, and Challengers* (Wiley, 2010)

V.P. Seidel, S.K. Fixson, Adopting design thinking in novice multidisciplinary teams: the application and limits of design methods and reflexive practices. J. Prod. Innov. Manag. **30**, 19–33 (2013)

L. Shostack, Designing services that deliver. Harv. Bus. Rev. **62**(1), 133–139 (1984)

M.A. West, C.S. Borrill, J.F. Dawson, F. Brodbeck, D.A. Shapiro, B. Haward, Leadership clarity and team innovation in health care. Leadersh. Q. **14**(4–5), 393–410 (2003)

Part II
Digital Thinking

Chapter 4
Introduction to Digital Thinking

4.1 Introduction

Digital thinking is referred to as an approach that involves the three essential aspects to address a real-world problem. These tasks are solving the identified real-world problems, designing the systems that can be employed by the society, and comprehending the human behavior that is based on fundamental concepts pertaining to the world of computing. It is often considered to be similar to analytical thinking in the sense that it establishes ways to address problems. It shares with design thinking the space to define the ways to solve real-life problems and to evaluate systems that are characteristic of being large and complex operating in the realm of real world. With terms to scientific thinking, this considers the processes that can aid in understanding computer intelligence and the mind-set of human behavior.

Abstraction is the essence to digital thinking. Notions are abstracted beyond the notions of physical dimensions associated with time and space. Human abstractions are perceived to be general whereas numeric abstractions are considered to be one of the special cases. Human abstractions are richer and complex in comparison with the physical and mathematical abstractions. Mathematical abstractions are clean and follow easily definable algebraic properties whereas human abstractions do not. As for instance, stack of elements are one of the most common types of data in the domain of computing. Humans will not consider adding up of two stack elements but will consider adding two integers instead. Another common example of abstraction is algorithm that takes certain input and produces some output. Programming language is another prominent example of abstraction encompassing set of strings. The interpretation of these strings results to effect the final computation. Combining two programming languages results into some other kind of abstraction that entails an entire research agenda to suitably define it. Secondly, human abstractions work under the realm of physical world and therefore one needs to take notice of edge and failure cases. These cases may be the problems arising as a result the disk gets

K. Kumar et al., *Design Thinking to Digital Thinking*,
Manufacturing and Surface Engineering,
https://doi.org/10.1007/978-3-030-31359-3_4

full or server not responding. The program encountered an error that should have otherwise been identified during the compilation process. Another edge case may be to avoid a programmed robot to ram into people when it moves down halfway. Therefore, defining right abstraction is critical to work with the rich abstractions. The abstraction process, i.e., what needs to be taken under consideration and what needs to be ignored is key to digital thinking.

Process of abstractions includes layers. At least two layers of abstractions are involved when working with computer environment. One is the layer of interest and the other is above or below it. Building of large complex systems is possible because of the well-defined interfaces between the layers. The unique characteristic feature of any application software and its component is that even if a user doesn't have the required knowledge and the details regarding the component in order to interact with the software component, the user interface makes it easier for the user to comprehend. Furthermore, the implementer of the software application may not be aware of the potential user of the software component so that the application is implemented correctly. As for instance, the architecture of Internet encompasses particular "thin waist" layer of Internet protocol and underpins at the bottom the inclusion of new software components and devices as well as technology associated with networking also the provision to entail any unforeseen application at the top. However, one of the things that require special attention is that related with relationship between different layers of abstraction. This could be established either by employing relations and equations associated with simulation, abstraction function or through the employability of a more generalized instances of mapping. The mappings are employed to show balance between machines in the abstract state to that of its possible refinements. Furthermore, mappings are also employed in ensuring whether the implementation is correct with respect to the specifications and in compilation of a program to more efficient machine code from initially written using high-level language.

In brief, the major basic elements of digital thinking include definition of abstractions, working with abstractions and the associated layers and understanding the relationships established between different abstraction layers. Therefore, abstractions can be considered as a critical tool of computing.

Digital thinking can be referred to as the automation of human abstractions. Humans operate by mechanization of abstractions, different abstraction layers, and hence the relationships between the different abstraction layers. Precise and exact models and notations make the process of mechanization possible. Automation implies that certain mechanism or devices are required to interpret the different abstractions. One of the most sophisticated devices is that of computer which is a physical device constituting of varied capabilities such as communication, storage, and processing. Machine can also be human as human process information and compute the associated information. In other words, the conceptual framework of digital thinking process does not require necessarily a machine. However, when human and computer are considered together, then the processing power of a human can be exploited and combined with that of the computer. As for instance, humans have better interpreting capabilities in comparison with the machines whereas machines execute certain instructions more quickly and precisely in comparison with the humans. Machines can also process larger data sets in comparison with human.

Therefore, digital thinking is mainly concerned with answering the pertinent question of how to involve a computer to get answer to a problem. Computer can be a human, a machine, or a combination of a human and a machine or a combination of such different computers. Therefore, to solve the problem it is quintessential to identify appropriate abstractions and hence choose a suitable machine that has essential power to solve the problem under consideration. Digital thinking therefore can offer more than what the simple usage of machines can provide.

4.2 Scope of Digital Thinking

Digital thinking has been influencing research propositions in nearly all the disciplines whether it being sciences or humanities. There are umpteen evidences to show the influence of digital thinking on other domains as well. As for instance, there has been transformation in the domain of statistics through digital thinking approach. It has become possible to decipher any errors in the voluminous data and information through the employability of machine learning and hence the automation of Bayesian methods. These data sets can be as adverse as grocery store receipts, credit card purchases, astronomical maps, and magnetic resonance imaging scans. Another instance is the transformation of biology underpinned by digital thinking. The ability to sequence the genome has been accelerated through the shotgun sequencing algorithm. This has also been extended to include human abstractions that being dynamic in nature (Fisher and Henzinger 2007). A new field of economics referred to as microeconomics has emerged as a result of integrating digital thinking with economics domain. The transformation in economics domain can be evidenced from umpteen applications such as online auctions, advertisement placement, and so on (Abraham et al. 2007).

There are, however, other domains wherein digital thinking has been still at the beginning stage, i.e., still number of machine cycles are used to solve the real-world problems. There is greater dependency of sciences and engineering on large quantum of computer simulations of different physical processes existing in the real-world and the associated mathematical models. Simulation of entire aircraft for space missions is one of the examples in aerospace domain. The geoscience on the other hand would be inclined to simulate the Earth to Sun. New trends as well as patterns can be established to understand the complex realm of humankind through the utilization of digital thinking methods such as data mining and federation. This has ultimately aided humanities to use the different resources such as artifacts and digital books to create such opportunities.

The future prospects of digital thinking are very promising and can be put to exploit more cleverer or sophisticated abstractions which may aid the scientific community as well as those in engineering to model their complex systems and analyze them on a greater magnitude than what is being handled in the present scenario. The use of abstractions layers can aid in modeling complex and large real systems at multiple time scales and modeling them at higher degree of resolution levels. The multitude interactions of the umpteen complex systems can be modeled in order to decipher the emerging behavior. The digital thinking can be used to increase the number of parameters as well as initial conditions associated with a real-world problem and hence provide a more realistic solution. The models can be modeled with respect to time and hence can be validated against the prevailing truth.

A more deep insight into digital thinking can aid in modeling more complex systems. The voluminous amount of data generated can be analyzed precisely. More data will be generated as a result of routine usage of surveillance systems, monitoring systems, deployment of distributed sensor nets, digitization of world's information, simulation results from complex systems, digital cameras on cell phones, and so on. It will be through the digital thinking approach the knowledge extraction process buried deeply will become more convenient. Through the mechanism of open feedback loop, the gained knowledge will ignite us to debate on relevant questions pertaining to the systems under consideration. This will result in fine-tuning the simulation models and thereby generating more data.

If digital thinking is employed by majority, then it will become ubiquitous gradually. This, however, raises pertinent question regarding education, i.e., if digital thinking is considered as one of the thinking abilities, then major challenge arises is to make arrangement for teaching such type of thinking approach to the students or other novices related to the field. Let us digital thinking is gradually taking place and influencing the skills of graduating students. Universities are incorporating the digital thinking approach in their curricula of graduate courses. They are able to perceive how the future generations will be able to think and expand their list of abilities that will aid them to succeed in the modern society. However, the very approach of digital thinking should be ensured among all levels of education systems. Such an approach can be ensured for all if learning process is undertaken during the early childhood days.

One of the pertinent questions that arise here is that related to the effective methodologies and approaches that can be employed to teach digital thinking to schoolchildren at different levels of education. To address this issue, one must be able to know the elemental concepts associated with digital thinking. Experts in the digital era have continuously answered to this question by creation of curricula system for undergraduates wherein the focus is not only on just teaching the skills associated with computer programming but also on the principles associated with computing.

The aforementioned question is pertinent to be discussed in collaboration with science and engineering students. Building blocks for digital thinking can be developed by incorporation of fundamental concepts of digital thinking into formal learning. Digital thinking can be approached if one considers the similar human closeness as to the mathematical concepts.

Another question relates to the precedence of teaching concepts to children as it is a well-known fact that their learning ability progresses gradually over the years.

As for instance, numbers are thought when the children are nearly five years old; algebra, when they become twelve years; and calculus, at the age of eighteen years. The progression can, however, be structured in number of ways which aids in effective development of digital thinking concepts and deciphering the other ways in which this thinking approach can be learnt.

Effective integration of digital thinking tools with the different teaching concepts raises another relevant question. The domain of digital thinking encompasses concepts related to computing as well as other tools such as the pertinent digital devices. The digital devices raise certain challenges as well as opportunities. One of the major challenges is the interference of tools in understanding the digital concepts. Experts in the field do not opine to let people use the tools without proper understanding of the concepts associated with a tool. Furthermore, the experts also opine that the users should not think that because they are adept to using a tool they have understood all the associated concepts. Another challenge that lies at the forefront is to track the usage of the tool in accordance with the concepts learnt. The point at which the powerful capabilities of computing machine must be introduced is another question of debate.

The concepts taught can be reinforced with the tools used and is therefore one of the arising opportunities. Digital tools are path that aids abstractions to come alive. As for instance, it becomes possible even at the early grades to effectively present visually the differences between a polynomial-time algorithm and an algorithm associated with exponential time. Or for another instance, it can be shown visually and hence comprehended that tree is a special kind of graph. In later grades, children will be able to automate their own abstractions using digital thinking approach. Hence, digital thinking approach not only aids in understanding the associated concepts but also aids in elucidating the concepts in other fields as well. Another opportunity that needs to be tapped is the routine exposure of children to the various digital devices both at the school as well as at the home. The exploration of formal as well as informal learning processes must also be taken care off. There are umpteen ways in which the learning process can take place outside the classroom such as they learn from their families, museums, hobbies, and life experiences.

A number of barriers such as social, cultural political and economic are associated with envisioning digital thinking to be integral part of childhood education. However, worthwhile benefits can be exploited if the digital thinking approach is integrated within the education framework.

4.3 Thinking About Digital Thinking

Technological innovations, scientific questions, and societal demands are the major elements that drive the digital thinking approach. A scientific question that is one of the bases of any field is often underestimated in the context of digital thinking process. Major focus on the societal demands and technological innovations are the major reasons behind this ignorance. Furthermore, the importance of digital thinking approach can be realized truly only if the importance of each one of the elements as well as their combination is realized which makes the domain of digital thinking distinctive.

There is an enchanting relationship of push and pull that can be realized between the aforementioned three driving elements. In one direction, technological innovations are fed by the newer scientific discoveries. This in turn feeds the societal applications. In the other direction of this relationship, new creative social uses are inspired by new technological innovations which in turn demand the attention of new scientific discovery. As for instance, society demands science is truly reflected in the following explanation: More and more computing as well as communication machinery is demanded to cater to the ever-growing mobile phones population and therefore demands new research and discoveries in sciences so that the energy related can be used more effectively. The following instance establishes the quick relationship between society and new technology. The existing network capabilities are being stressed owing to the demand for higher fidelity and realistic environment. The network facility is stressed to transmit real-time multiple multimedia simultaneously with the required confidentiality. The fundamental desire to connect and communicate with the like-minded people has resulted in rapid increase of various social networking platforms such as YouTube, Twitter, and Facebook. This desire in turn has added another dimension to the rapidly growing economy.

Moore's law is doomed to its end in the near future with the beginning of digital era. As a consequence of this beginning, the silicon-based technology is challenged with the production of machines that have multi-core ability and to program effectively the parallel processing capabilities. Various prospects of nanocomputing, quantum computing, and even bio-computing are now being explored. Although many of the associated technologies have arrived. As for instance, seven-point Hamiltonian path problem has been addressed by Adleman (1994) with DNA computing. IBM (2006) has declared the development of integrated circuit on a carbon nanotube molecule. Molecular machines are now being developed as consequences of rapidly changing technological propositions. Swiss have employed quantum cryptography to secure their ballots.

Memristor (Chua 1971) has been announced to be developed by Strukov et al. (2008). Memristor has been referred to as a missing fourth element between inductor, capacitor, and resistor. The usage of mobile phones, robots, sensors, and actuators has grown exponentially. The automobiles are now laced with network of computers. As far as quantum of data is concerned, it has become voluminous. This is because of availability of cheap data storage facilities, ubiquitous presence of sensors. Owing to the bulkiness associated with data, the current population is experiencing information overload. Web 3.0 is one of the invigorating topics of research in the domain of communication. It has become possible to visualize a more sophisticated world. This potential ability will facilitate the scientists as well as the engineers to communicate and conduct their work in tandem with scientists over the side of the globe. This is facilitated through the development of virtual organizations.

Scientific community now aspires to develop machineries that mimic the human brains. As for instance, Blue Brain Project carried in collaboration by IBM with EPFL

(2005) has successfully created model of brain that carries out its functionalities as well as is biologically accurate. Another instance is that of a start-up company known as Numenta (2005) that has successfully created software platforms for intelligent computing that mimics the human limbic system. These are some of the interesting trends of digital thinking approach and it will be one of the reasons for rousing curiosity in the domain of digital thinking in the coming years.

The technological advancements in the realm of information technology have raised the expectations of the society. People across the globe now demand for solutions that are cent percent reliable and are available to them as and when desired. They demand for cent percent connectivity and availability. With the advancements in computers and communications, people are now demanding for instantaneous response, storage facilities that can store anything forever and also the accessibility of stored data from anywhere. There is variation in the class of technology users and is not only limited to engineers and scientists but also includes illiterates and general literates that are both young and old, rich and poor as well as abled and disabled.

Technologies must also have the potential ability to cater to the aforementioned classes of users. These users can be either individuals or in population groups. Companies are now able to track the users' queries and therefore personalize the devices that are being used by the society. The advertisements have been personalized as the search companies are now able to keep track of what we see. Information and technology is being used to the extent that society is now using them to preserve their cultural heritage. Networking platforms such as LinkedIn and Facebook allow users to connect with the like-minded and share information. However with advancements come associated challenges as well. This may include accountability, anonymity, privacy, and identity management. Another challenge is to balance openness with privacy.

Scientific challenges have been revealed to be associated with research pursuits and technological innovations (Wing 2008a, b). There are number of scientific questions pertaining to the computer world and hence the digital thinking horizon. As for instance, the questions associated with intelligence, computability, information, and the reliability with which the complex digital systems can be built.

4.4 Conclusion

Digital thinking has become one of the key requirements for current as well as future engineers. Digital thinking has in fact aided design thinking approach wherein scientific community as well as young engineers can test and analyze their models before their final implementation. Important competencies can be developed through the inclusion of digital thinking approaches (García-Peñalvo 2018). Facilities to store data economically have been made available by the continual progress in the field of computer and information technology. With advancements in technology, societal demands are also on the rise. The anonymity, privacy of data has become one of the challenging aspects of digital thinking process. It will be interesting to track the advancements that can address the aforementioned challenges and therefore result in true realization of digital world through digital thinking approach.

References

D.J. Abraham, A. Blum, T. Sandholm, Clearing algorithms for barter exchange markets: enabling nationwide kidney exchanges, in *Proceedings of the 8th ACM Conference on Electronic Commerce*, June 2007 (ACM), pp. 295–304

L.M. Adleman, Molecular computation of solutions to combinatorial problems. Science **266**(5187), 1021–1024 (1994)

L. Chua, Memristor—the missing circuit element. IEEE Trans. Circuit Theory **18**(5), 507–519 (1971)

J. Fisher, T.A. Henzinger, Executable cell biology. Nat. Biotechnol. **25**(11), 1239 (2007)

F.J. García-Peñalvo, Editorial computational thinking. IEEE Rev. Iberoam. Tecnol. Aprendizaje **13**(1), 17–19 (2018)

D.B. Strukov, G.S. Snider, D.R. Stewart, R.S. Williams, The missing memristor found. Nature **453**(7191), 80 (2008)

J.M. Wing, Computational thinking and thinking about computing. Philos. Trans. R. Soc. A Math. Phys. Eng. Sci. **366**(1881), 3717–3725 (2008a)

J.M. Wing, Five deep questions in computing. Commun. ACM **51**(1), 58 (2008b)

Chapter 5
Digital Thinking in Education

5.1 Introduction

It is one of the difficult tasks to suitably delineate the concepts and the conceptual framework associated with the digital thinking approach. This is owing to the fact that many of the studies in this domain use programming as their major context and this can be very confusing to those in the initial stage of grasping the conceptual framework of digital thinking as different studies form opinion that digital thinking is same as programming or to say the least digital thinking requires programming. The confusion is further aggravated because of the associated history that the thinking skills are developed through the usage of programming languages. However, digital thinking does not entail the employability of programming necessarily and nor there are claims from the experts in the domain of digital thinking that opines for programming to be the basis for development of the related skills. A proper examination of the research articles regarding development of thinking skills and programming will result into understanding as to why digital thinking has undertaken this alternate pathway in development of thinking skills.

The relationship between programming and thinking skills has been appropriately delineated in a book from Papert (1980). The book referred to the effect that LOGO programming language has on the learning abilities of student and the concepts associated across multiple disciplines. It was revealed that while on the one hand the prime focus of LOGO programming language was on explicit learning of programming, digital thinking on the other hand employs the general concepts associated with computer science domain. To most of the surveyors in the domain, only slight distinction between the two might be observable. They opine that it is only through the basic knowledge and understanding of the concepts of computer science domain, the foundation of digital thinking approach was developed. Game design is one of the popular approaches to teach programming skills (Chiazzese et al. 2017).

K. Kumar et al., *Design Thinking to Digital Thinking*,
Manufacturing and Surface Engineering,
https://doi.org/10.1007/978-3-030-31359-3_5

Papert (1980) opined that the students would gradually develop themselves into epistemologists through the usage of LOGO programming language. Therefore, it was thought that through the usage of this technology the skills would be developed that would motivate students to think about their thinking capabilities. On the other hand, Pea et al. (1985) created tasks associated with the daily classroom activities. The tasks were designed so that the knowledge grasped during their training with LOGO could be well used in conduction of the assigned tasks. The students involved were provided training for a year. However, the investigation could not decipher that the planning skills in performance of no-programming activities possessed by students with experience on LOGO were much better in comparison with that without any experience on LOGO. The transfer of debugging skills to non-programming activities was however reported in students with training on debugging through LOGO. As for instance in rectification of the problem associated with planning travel route, students that were experienced in using LOGO were effectively able to find bugs in the process in comparison with the students that did not have any LOGO experience. The effect of learning lasted even after few months of finished courseware on LOGO instructions.

The aforementioned literature clearly reflects on the distinction in findings. The possible explanation of the revealed differences was given by Salomon and Perkins (1989). This was done through their examination of high- and low-road transfers. The skills that formed part of low-road transfers were the ones that were repeatedly practiced. The number of times the task was practiced and the frequency with which it is practiced in a particular context revealed the amount of transfer. While on the other hand, mindful abstractions of the processes or concepts being grasped encompassed high-road transfer. A distinction can be made between low-road and high-road transfers on the basis of their reflexive nature as high-road transfer is beyond to low-road transfers in this context. Quantum of experience was demanded for transfer of low-road programming skills to solve non-programming problems. On the other hand, high-load transfers were reported to be accomplished through only careful instructions on applying the grasped programming skills to any contextual framework. It is the experience of the educator on the instructional practices that decided the efficiency of high-load transfer. This leads to an important question at this juncture of discussion, i.e., what really makes the difference: Programming experience or the methodology adopted by instructors. Therefore, it is clear that programming alone is not enough for learning and programming is just like other skills that are taught only in one context. Digital thinking has been used as a key driving force behind programming skills (Tedre 2017).

The above findings can aid in effective conceptualization of digital thinking process. It can be defined in one manner as the teaching of thinking skills that can be employed to address the problems across disciplines. One of the major forces to teach skills associated with mental abilities is that of effective programming instructions. Learning to program can result in enhanced ability to solve problems, academic outcomes and grasp thinking skills. Therefore. digital thinking has its foundation in programming that aids in learning of associated and desired mental skills. Digital thinking often asks pertinent question as to why not learn the thinking skills that can have widespread applicability, i.e., can be applied to multiple domains. However, experts opine that focus on instructional methods and general skills can be a more effective solution for ensuring effective learning of thinking skills and not only the self-centered approach on programming alone.

Programming and digital thinking are often confused with one another but are not equivalent conceptual frameworks. Computing and the related programming is one of the set of skills that drives the digital thinking approach. Digital thinking arose from the domain of computer science and is not only the discipline wherein one can use these skills of thinking. Therefore, it will be a mistake both pedagogically and conceptually to expect programming to result into efficient thinking approaches. The primary focus should be on learning high-level programming and computing concepts and the multiple disciplines in which the learnt skills should be applied.

5.2 Defining Digital Thinking and the Associated Challenges

The major idea behind defining digital thinking is itself challenging. The challenge mainly lies into differentiating between the core and peripheral qualities of digital thinking approach. This approach of defining digital thinking aids in identification of set of necessary conditions that are required to be met for by certain practices to be regarded as digital thinking skills. However, it is not always possible to go for the aforementioned approach of defining the cognitive processes owing to the difficulty in effective implementation.

As for instance, few of the experts argue on the number of attitudes and the essential dimensions for digital thinking processes. Some of the argued attitudes and dimensions include confidence, patience to work with the complex problems, confidence to address complexity, tolerance and resistance to deal with open-ended real-world complex and large-scale problems and the potential ability and the capacity to work and coordinate in a team environment. However, the above list of dimensions is worth in consideration to conceptualizing digital thinking. However, expansion of the list can dilute the very foundation of digital thinking approach. The distinction from other skills of the twenty-first century is blurred with the proposition of expanding the list. Another aspect is that if the expanded list of attributes is concerned then the focus will be on logical definition and not on the pragmatic approach.

Therefore, a more practical approach in defining digital thinking approach is quintessential because of the flexibility that the approach does not require understanding certain set of associated concepts in development to series of necessary and essential dimensions and attitudes required to be met for skills to qualify for digital thinking. A more general graded notion that encompasses the possible rather than necessary proposition in defining digital thinking is preferred.

It is better to search for similarities, differences, and relationships between elements that possess overlapping characteristics and sometimes similarities and differences of detail in order to tread the path of pragmatic approach. Tersely, it is better to undertake a broader and philosophical view that aids in providing us with current thinking approach in the domain of cognitive sciences and therefore a bird's eye view to identify and encompass peripheral skills in definition to digital thinking.

Even after the associated challenges, a number of definitions have been provided by the scholars and academicians across the globe. Design thinking complements thinking prospects in engineering and mathematics domain with focus on designing systems and solving complex problems faced by humans in real life (Wing 2008a, b). The core of digital thinking approach encompasses abstractions, layers, and relationships between abstractions and layers. Abstractions signify the mental tools desired to solve the problems and layers are significant that a problem consists of different levels. The fundamental to digital thinking is the concept and idea of abstractions and the ability of students to judiciously deal with different abstraction layers. In addition, it also includes thinking algorithmically and understanding the consequences of big data (Denning 2009). Further, it has been argued that digital thinking comprises of formulation of problem and the solutions involve computational steps and algorithms. Algorithmic thinking was therefore the connotation for the digital thinking approach since it involved conversion of input to outputs and therefore in doing so certain algorithms were required. Understanding of concepts associated with programming is not essential to comprehend the conceptual framework of digital thinking. A total concentration on programming skills can hamper student progress in their discipline of computer science. Digital thinking can therefore be defined as a skill that can process information systematically, efficiently, and correctly. The same applies to tasks as well which ultimately aids in solving complex problems.

Digital thinking has been considered as an important twenty-first-century skill set by many in the field of education and in particularly the education community of the computer science discipline. Several definitions have been given birth for importance of digital thinking and education owing to the numerous definitions and concepts associated with design thinking. The key elements in these definitions are the focus on skills, desired attitude for complex problem-solving with the employability of computers and habits (Grover and Pea 2013; Sengupta et al. 2013; Lee et al. 2011). Digital thinking encompasses to possess innate ability to solve complex problems by dividing abstractions into several levels through application of design-based reasoning and mathematical reasoning. It has been often argued that digital thinking approach often goes beyond normal human–computer interaction. Digital thinking could motivate students to create new tools and hence foster creativity. Therefore, students do not remain to be consumer of knowledge only.

5.3 Digital Thinking and Education

Experts opine that digital thinking should be considered a universal skill set that should be made integral part of students' curriculum with not only computer science background but also from other disciplines. It is important competency ability that influences nearly all the domains of education. Ability to think digitally forms the basis of conceptual understanding in every educational domain as it involves the process of problem-solving and algorithmic thinking. Since in the present scenario, students are required to deal with computing in their everyday life; hence, digital thinking cannot be ignored as an important skill set for desired competency. Owing to the importance of digital thinking, experts underpin the inclusion of digital thinking into the educational curriculum and therefore offer students the desired industrial skill set to address real-life complex problems.

However, the basic question arises as to what should be taught. Although the importance of digital thinking is well realized, the major challenge lies in regarding the content and the timing of the course to be taught. The course must not be made of only for the undergraduate students and those at the universities but should also be integrated with all the levels ranging elementary to higher school. As far as learning time is concerned, the learning should be accomplished during the early childhood years of life so that strong foundation of concepts and core associated with digital thinking can be achieved. Another important question that rises at this juncture of discussion is as to what relevant concepts must be taught at different stages of schooling. The abstract concepts that are taught through animations and visualizations could be reinforced through digital thinking conceptual framework. Digital thinking can be categorized into nine core concepts, i.e., collection of data, analysis of data, representation of the data, decomposition of problem, abstractions, algorithms and procedures, parallelization, automation, and simulation. There are several characteristics that could be realized in considering digital thinking as a key to solve fundamental and complex problems. Some of these can be listed as follows: formulation of problem in such a way that the problem could be solved using computers and other relevant tools, organization of the data logically and their logical interpretation, representation of data through simulations and models, employability of algorithmic thinking to solve identified problems automatically, identification and analysis of possible solutions with the sole objective of achieving the effective and efficient combination of steps and ultimately generalizing the identified solution so that the implementation can be done to address wide variety of problems. The aforementioned characteristics separate the cognitive process of digital thinking to that of solely working on a computer and other digital devices.

It is also quintessential to factor in various issues of human intuition, expertise, and knowledge when discussing on digital thinking process. The success in employability of digital thinking concepts is not solely dependent on the knowledge acquired in the domain of computer science and mathematics but also on the other cognitive and imaginative thinking skill sets and capacities such as design thinking and innovative thinking.

5.4 Digital Thinking into Curriculum

One of the important aspects in integrating digital thinking with the educational curriculum is judiciously defining the boundaries with other domains and disciplines and other competencies related to the twenty-first century. Digital thinking can be positioned in the curriculum of computing with the objective that every student should be provided the chance and opportunity to learn and grasp the concepts associated with computing from the early childhood days and by the age of fourteen should be able to work in a specified qualification area. A detailed report on integration of digital thinking within education curricula should be carried out by the economies over the world so that implementation of digital thinking can be effectively commenced with due consideration to future prospects and requirements in the market.

Experts have opined to identify the major skills and the thinking capabilities in the realm of computer science and to teach these identified learning sets into other contexts. There has also been debate pertaining to linking of digital thinking approach to other subjects apart from computer science. The main aim of digital thinking approach is to allow its users not to feel just if they are computer scientists but also as economists, artist and to comprehend the ways in which complex problems can be solved. Furthermore, they should be able to acquire qualities that can aid them in creating and discovering new questions that can be explored to the best limit possible. It allows for development of new ways of thinking that can allow users to allow the learners to explore new digital thinking tools in creative ways within and across the different disciplines. Hence, it comprehends the boundaries of digital thinking within other disciplines that requires greater attention from the scientific community.

Number of instances and frameworks has been provided by researchers and educators for incorporation of digital thinking approach across different disciplines. The concepts and skills associated with mathematics, computer science, social science, languages, and arts form the core conceptual framework of digital thinking approach. The integration of digital thinking with different subjects will result in increased employability of terms and concepts associated with computational domain and acceptance of attempts that have led to failed solutions. The teamwork environment should be motivated simultaneously so that the important aspects of digital thinking approach, i.e., decomposition, abstraction, negotiation, and building of consensus can be practiced appropriately. Digital thinking and the associated notation can be employed in mathematics as for teaching and learning multiplication tables, finding square roots, and charting information. It could be employed in social science domain to comprehend concepts such as that related to assembly line. Learning and teaching grammar can be exploited in the domain of language-arts. Algorithms, engineering concepts and principles, scientific inquiry can be exploited and applied in programming-based environment that is based on agents. The students therefore can use these to accomplish their task of designing, evaluating, refining their models across the disciplines. To adjudge the effectiveness of digital thinking approach, many have integrated the digital thinking approach across critical domains. Digital thinking approach has been integrated in elementary schools in developed nations and efforts have commenced in this direction in developing nations. Robotics for instance has been integrated with digital thinking process and has fostered the thinking approach in young children (Bers et al. 2014). The concepts of digital thinking process were linked to concepts from literacy and mathematics.

Six different phases associated with a digital thinking approach should be identifiable when integrated with any curriculum: computing, development of computational artifacts, abstracting, analysis of problems and developed artifacts, communicating and collaborating. There have been significant impacts of computing on the society as it has led to a number of discoveries and innovations. These technological developments have huge impact and implications for commercial market, individuals involved and hence society as a whole. The students in an integrated course will get acknowledged of these connections and hence learn and comprehend to draw connections or the relationships between the various concepts of computing world. Once understood, students must be able to design and develop computational models and the associated artifacts. The learnt techniques and concepts are employed by students to solve problems. Hence, in this phase of the curriculum students are expected to develop artifacts through selection of appropriate techniques learnt and use appropriate algorithms and other principles of management. The process of digital thinking motivates students to understand and apply abstractions at different levels such as that from privacy in social networking space to human genome project. Therefore, students in the third phase of curriculum must be able to develop models, simulate artificial as well as natural phenomenon and hence use these simulations to predict and analyze their efficacy and validity. In the fourth phase of curriculum, students must be able to comprehend the results and the developed computational artifacts through the application of different criteria that are pragmatic, mathematical, and aesthetical. Students will be able to produce their own solutions and analyze their computational work that has been produced by them. Through analysis, students must be able to locate and correct errors and hence justify their appropriateness and correctness. In the fifth phase of curriculum, students are able to describe computation and justify their choices through the aid of graphs, computational analysis and graphs. Hence, students must be able to explain the meaning of result and summarize the sole purpose behind the generated artifacts. In the last phase of the curriculum, students are expected to learn to work in a collaborative environment. They will collaborate in activities and hence investigate questions and associated data sets.

5.5 Digital Thinking and Informal Learning Space

There have been questions on the need to position digital thinking approach in the school space. It is always better to associate digital thinking to diverse learners that encompasses even the learners that are engaged informally. The learning process also takes place beyond the classroom space, i.e., in informal settings too. Children are not afraid to explore new things and explore them. They can learn from each other, from their parents, their family, learn through hobbies and other life experiences.

Digital thinking environment can be provided even in the context of game design, modeling, and simulations. This can be accomplished through a use-modify-create pattern.

5.6 Conclusion

The idea of digital thinking is a thinking ability and skill set that must be possessed by children since their early years. The importance of digital thinking has grown exponentially since its inception. Process of digital thinking encompasses concepts associated with the domain of computing world but the two fields are not similar in different perspectives. Digital thinking entails the thought process to solve complex and large-scale real-world problems and therefore generalizing the ability to solve wider range of problems disbursed across different disciplines. Through the process of digital thinking, the computational world could be understood well and aid into reason about the various natural and artificial processes and systems of nature.

However, the major challenge lies into define the process of digital thinking that has to do with suitably defining the core abilities and competencies as well as the peripheral skill sets associated with the digital thinking process. Furthermore, the effectiveness of digital thinking process can only be realized through its integration with the educational curriculum. It is challenging to prepare teachers to effectively implement the integrated curriculum. Teachers and educators from across the disciplines need to get acquainted with the conceptual framework core to the digital thinking approach. It is required to include real-life day-to-day examples in order to comprehend the concepts associated with the digital thinking process. Various projects worldwide have been introduced to provide the associated competencies to both the teachers and students (Pinto-Llorente et al. 2018; Sáez-López et al. 2016).

The research is very scarce in integration of digital thinking process within educational framework and hence there is a requirement in the present scenario to study and investigate how digital thinking can be established in students from different backgrounds apart from that of computer science background.

References

M.U. Bers, L. Flannery, E.R. Kazakoff, A. Sullivan, Computational thinking and tinkering: exploration of an early childhood robotics curriculum. Comput. Educ. **72**, 145–157 (2014)

G. Chiazzese, G. Fulantelli, V. Pipitone, D. Taibi, Promoting computational thinking and creativeness in primary school children, in *Proceedings of the 5th International Conference on Technological Ecosystems for Enhancing Multiculturality*, Oct 2017 (ACM), p. 6

P.J. Denning, Beyond computational thinking. Commun. ACM **52**(6), 28–30 (2009)

S. Grover, R. Pea, Computational thinking in K-12: a review of the state of the field. Educ. Res. **42**(1), 38–43 (2013)

I. Lee, F. Martin, J. Denner, B. Coulter, W. Allan, J. Erickson, L. Werner, Computational thinking for youth in practice. ACM Inroads **2**(1), 32–37 (2011)

A.M. Pinto-Llorente, S. Casillas-Martín, M. Cabezas-González, F.J. García-Peñalvo, Building, coding and programming 3D models via a visual programming environment. Qual. Quant. **52**(6), 2455–2468 (2018)

J.M. Sáez-López, M. Román-González, E. Vázquez-Cano, Visual programming languages integrated across the curriculum in elementary school: a two year case study using "Scratch" in five schools. Comput. Educ. **97**, 129–141 (2016)

P. Sengupta, J.S. Kinnebrew, S. Basu, G. Biswas, D. Clark, Integrating computational thinking with K-12 science education using agent-based computation: a theoretical framework. Educ. Inf. Technol. **18**(2), 351–380 (2013)

M. Tedre, Many paths to computational thinking. Paper presented at the TACCLE 3 final conference, Brussels, Belgium, 2017

J.M. Wing, Computational thinking and thinking about computing. Philos. Trans. R. Soc. A Math. Phys. Eng. Sci. **366**(1881), 3717–3725 (2008a)

J.M. Wing, Five deep questions in computing. Commun. ACM **51**(1), 58 (2008b)

Chapter 6
Digital Thinking Integrated with Design Thinking

6.1 Introduction

Many in the design space has been opining for development of relevant design methods and tools in the changing world scenario that is plagued with challenges and complex contemporary issues. In the similar time frame, there are other group of individuals that opine for digital tools to address the challenges associated with pedagogy, technology, innovation, and communication. The first group is known to be proponent of design thinking while the second group is proponent of digital thinking. On the one hand, both the proponents of design thinking and digital thinking motivate to promote abilities such as constructive thinking, collaboration and multidimensionality, the proponents of digital thinking support the employability of digital technologies as for instance computers, mobile devices, Internet and social media to address the challenges associated with the higher education. However, it has been believed by the design thinking community that focus on technology and digital can suppress to an extent the role of design thinking capabilities.

The digital thinking approach has been instrumental in reconsideration and improvement of the existing educational models. Literacy skills associated with code development have been considered as crux of science, technology, engineering, and mathematics (STEM) subjects (Weintrop et al. 2016). The reconsiderations can be undertaken at different levels of education that ranges from individual classrooms to enterprise-level decisions wherein the entire academic infrastructure is reconfigured to accommodate digital infrastructure. The digital thinking process is spearheaded by number of diverse fields ranging from literature to computer science domain. The driving force is the realization that the world is very different than what existed decade ago and also the requirements of student community have now been changing drastically. Therefore, in short the digital thinking approach focuses on the following: the required competencies by the citizens of the twenty-first century globally that are existent in the digital societies, rethinking of the existing notion of literacy that can aid in inching toward the required competencies, the ways that should be

K. Kumar et al., *Design Thinking to Digital Thinking*,
Manufacturing and Surface Engineering,
https://doi.org/10.1007/978-3-030-31359-3_6

adopted to account for any paradigm shift and the new models for research, student product models and the scholarships required.

Design thinking comes into play here where the rethinking and reconsideration open their doors to the enduring approaches which can be accomplished through the encompassed design methodologies. Therefore, it is often encouraging to encompass both the thinking approach to address the aforementioned challenges. The present chapter aims to present an expanded overview of the design thinking framework that is built basically on the foundation of digital thinking approach entailing the expertise and experience of communication and interaction design approach.

6.2 Digital Thinking and Design Thinking

Scholars have referred the young people using computers, Internet, and cell phones as digital natives while there are others who oppose the usage of the aforementioned term (Jenkins 2007). The present connotation differentiates between the younger generations and the older ones. However, when perceived from view of social practices, cognition attitude, and learning styles, the aforementioned theme differs genuinely from that prevalent today. The present-day generation constantly engages themselves with the different enabling technologies. Hence, the digital thinking approach calls for reconsideration of existing pedagogy so that the learners with abilities to accumulate information from different resources and have strong visualization skills apart from being intuitive communicators.

The different initiatives promoting digital thinking and learning approach have been prevalent with the developed nations who are determined to provide the skill set required for the twenty-first century. In such cases, the advocates come from the technology field or have digital abilities. There have been initiatives from the different organizations that support different instructors, programs, and even the entire programs. The learnings as a result of such support not only remain enclosed within the classrooms but also extend to be able to access information and hence contextualize the accumulated information so that it can be employed within its peers (Ito et al. 2009).

Irrespective of the attributes that are required for generation to be called digital, a generation is also required to understand the different modes through which the competencies can be established and gained. Alongside the modes and tools for digital literacies, there have been calls for innovations in the domain of multimedia and digital literacies which could be accomplished through the design thinking approach.

6.3 Design Thinking in Twenty-First Century

Design thinking aids in identification of methods that can be used throughout different levels of educational system. In particular, it closely aligns with the needs and attributes identified by the advocates of digital thinking. Abduction as one of the approaches may be employed to cater to the daily needs in a situation where in the world is encountered with rare supplies of the required information. Moreover, abduction can function at its best by making something and reflecting on what is being made. Therefore, there is transformation to analytic component from intuitive aspect of the design thinking approach.

Approaches of educational domain related to digital learning process tend to value the process of design thinking which is concrete and promotes for practical experiences. The design thinking approach incepts from a number of incomplete evidences and then builds constructively to build toward a more abstract conceptual outcome. The proponents of design thinking believe that a large amount of cognitive abilities of human have been overlooked in educational system and the traditional educational system is no longer advantageous to the present existing system. Design fills this gap through inclusion of cognitive thinking abilities and hence makes it an important component of educational system apart from making it foundational component.

A connective framework between design thinking and digital think could be established when looking at the expertise of communication and interface. This could ignite a possible frontier open to research in this direction. An expanded notion of the design thinking approach can be adjudged from the following conceptual frameworks identified from specific practices and models tested time again and again (Burdick and Wills 2011):

User-oriented: A technical shift from a technical orientation-based tool to approach that supports user-orientation can be thought of owing to the interactive design perspective. This aids in designing of digital skills, tools, and media that takes into consideration the kinds of subject positions as well as the manner in which the application is employed. The views of the world and the associated models to a subject can also be taken into account (Moggridge and Atkinson 2007). This practice that entails user-orientation can also be combined with learner-centered theories of education.

Situated and networked: A universal frame of reference can be employed for construction of new models through the usage of designer's practical experience with the systemic embeddedness and dependence on contextual elements. It is the center of Gui Bonsiepe's diagram of ontology that encompasses the interface. Generic notion of literacy as a general skill set can be replaced through the aid of interdependent condition that altogether defines the interface. This allows for realization of literacy as an imagined literacy that is networked and situated.

Interpretive and performative: a designer's communication and vocational training typically includes certain aspects associated with visual rhetoric and narrative. This is realizable especially in the domain of field that is primarily concerned with making meaning. These models aid in construction of models for design thinking approach and derive their roots mainly from communication theory. These models aiding in establishment of notion of design thinking also accounts for different dimensions of media particularly the symbolic dimension. These can be linked directly to the notion of digital thinking as one can comprehend interpretation as critical as any critical practice.

6.4 Design Thinking Integrated Digital Thinking in Curriculum

Successful education system for digital thinkers tends to include the aforementioned qualities of design thinking. Number of schools in the developed nations as for instance "Quest to Learn" a school in New York City provides an example wherein the two thinking approaches, i.e., digital and design thinking coalesce with each other. In such schools, principles associates with design of game promoting discovery, innovation and strategic thinking forms the core of the educational curriculum.

The different schools with integrated design and digital thinking approaches employ design methods as tool for teaching. Game design aids in the development of abduction reasoning. A higher degree of relevance could be accorded to the educational system through the aid of iteration and problem-solving capabilities. As a result of changing technology, the educators and hence the students are required to learn new technologies. It has become one of the important requirements to constantly learn as to how to solve new problems every day. Game design allows students to engage themselves with subjects deeply, develop the required content that suits the present context and to develop the skills that can promote collaborations and therefore solve complex problems and ultimately design the products.

The role of design thinking could be established through the situated nature of education that encompasses to take into account the user interface. Furthermore, the employability of the practices associated with the design methodologies can also play a prominent role in establishing the design thinking contextual framework.

Even at the college level, one could well perceive the rise of digital thinking tools that has motivated the educators to take into account large effect of the approach of design thinking process across the educational curriculum framework. Balsamo (2009) has acknowledged the design thinking process and its activities as an important element to be incorporated into basic programs of educational systems.

The reconsidered educational framework developed by different educators has been tested in various universities. As for instance, the program crafted by Balsamo was intended for the students a University of Southern California. The main idea behind the program was to motivate the students to adopt new digital and design tools. The best possible place for the new program was in the general education curriculum which affected almost every student of the university. The courses associated with industries were integrated with the multimedia labs at the university. These were not only meant for the students to learn to employ software applications or make presentation but were also to realize that the outcomes and the reconsidered pedagogical approach relied heavily on the design thinking process. It was proved that the situated nature of the design thinking process proved to be a critical element for redesigning of curriculum for the new digital natives.

Although the coalescence was successful, the designed courseware had very low enrollment and hence redesign of the program was undertaken. Students were reported to resist the extra work which they found to be of little connect. Hence, the program was reconsidered to not include the introductory courseware to everyone but be provided as standalone experience. The students are provided with more support at higher level of courses to employ their research through the design and digital thinking skills.

Educators have realized that the new generation of students within the digital era requires new ways of thinking that associates with the design and digital environment. The new ways of thinking and expanded notion of such thinking approach should consider the following: a user-oriented approach, comprehension of situated expression, and visual rhetoric.

It is quintessential for students to learn the integrative thinking approach as this may help the scholars across the discipline to understand the different scholarly capabilities and the associated practices. Indeed, the scholars have responded positively to the new digital methods and technology positively. This has motivated the educators to build and develop on cyber-infrastructures. As for instance, the courseware is now available in number of digital formats such as videos, networked writing, and other architectures associated with the IT infrastructure. The scholars have realized the importance of new teaching approaches through wide range applicability of integrated design and digital thinking approaches.

Scientific journals have adopted the new digital paradigms as seen through the digital developments taking place across the disciplines. As for instance, the growing instances of digital humanities have been reflected in different literature (Hockey 2004; Drucker 2009).

It is clear that in the digital era there is dire need for innovative projects. Conceptualizing the dynamical and digital scholar scenario will remain to be a topic of debate among the educator community. This will be realized with the incorporation of adductive reasoning into the education domain. Understanding of the users and their attitudes toward digital and design platforms is critical for effective application of integrative thinking approach. This provides an open opportunity for practitioners to investigate and analyze the role of different design as well as digital platforms in the representation of knowledge.

However, designers are required to act urgently and lock in the design model or concepts before the scholars fully learn to access their own developed models and incept to integrate with the learning platforms. The sense of urgency has been acknowledged by the experts in the domain of educational curriculum.

6.5 Conclusion

At this critical juncture of educational transformation, the role of design thinking has to be realized in the digital thinking approach. It will be worth to question the role of design in educational system. Design thinking and digital thinking experts have valuable contributions toward the educational projects that have been produced through collaborative efforts from the classrooms and the knowledge skill set production houses. Design thinking integrated with digital thinking approach will aid in comprehending the world in digital era from the perspectives of the future generations. Innovations can be established in educational sectors that can have a long-lasting impact on the digital society as a whole. Other integrated platforms such as integrating the two design thinking approaches with sustainable development can be explored in near future (Gould et al. 2019). This also comes associated with its own set of challenges (Dekoninck et al. 2016). However, still many efforts are required to induce and implement the reconsidered educational framework.

References

A. Burdick, H. Willis, Digital learning, digital scholarship and design thinking. Des. Stud. **32**(6), 546–556 (2011)

E.A. Dekoninck, L. Domingo, J.A. O'Hare, D.C.A. Pigosso, T. Reyes, N. Troussier, Defining the challenges for ecodesign implementation in companies: development and consolidation of a framework. J. Clean. Prod. **135**, 410–425 (2016). https://doi.org/10.1016/j.jclepro.2016.06.045

J. Drucker, *SpecLab: Digital Aesthetics and Projects in Speculative Computing* (University of Chicago Press, 2009)

R.K. Gould, C. Bratt, P.L. Mesquita, G.I. Broman, Integrating sustainable development and design-thinking-based product design, in *Technologies and Eco-innovation Towards Sustainability I* (Springer, Singapore, 2019), pp. 245–259

S. Hockey, The history of humanities computing, in *A Companion to Digital Humanities* (2004), pp. 3–19

M. Ito, S. Baumer, M. Bittanti, R. Cody, B.H. Stephenson, H.A. Horst, D. Perkel, *Hanging Out, Messing Around, and Geeking Out: Kids Living and Learning with New Media* (MIT Press, 2009)

H. Jenkins, Reconsidering digital immigrants (2007), http://www.henryjenkins.org/2007/12/reconsidering_digital_immigran.html

B. Moggridge, B. Atkinson, *Designing Interactions*, vol. 17 (MIT Press, Cambridge, MA, 2007)

D. Weintrop, E. Beheshti, M. Horn, K. Orton, K. Jona, L. Trouille, U. Wilensky, Defining computational thinking for mathematics and science classrooms. J. Sci. Educ. Technol. **25**(1), 127–147 (2016)

Bibliography

A.M. Agogino, S. Sheppard, A. Oladipupo, Making connections to engineering during the first two years, in *Frontiers in Education, 1992. Proceedings. Twenty-Second Annual Conference*, Nov 1992 (IEEE), pp. 563–569

C.J. Atman, R.S. Adams, J. Turns, Using multiple methods to evaluate a freshmen design course, in *30th Annual Frontiers in Education Conference. Building on a Century of Progress in Engineering Education. Conference Proceedings*, IEEE Cat. No. 00CH37135, vol. 2 (IEEE, 2000), pp. S1A–6

E. Christophersen, P.S. Coupe, R.J. Lenschow, J. Townson, *Evaluation of Civil and Construction Engineering Education in Denmark* (Centre for Quality Assurance and Evaluation of Higher Education in Denmark, Copenhagen, Denmark, 1994)

R.N. Cortright, H.L. Collins, S.E. DiCarlo, Peer instruction enhanced meaningful learning: ability to solve novel problems. Adv. Physiol. Educ. **29**(2), 107–111 (2005)

A. Dong, A.W. Hill, A.M. Agogino, A document analysis method for characterizing design team performance. J. Mech. Des. **126**(3), 378–385 (2004)

D.L. Evans, Design in engineering education: past views of future directions. Eng. Educ. **80**(5), 517–522 (1990)

D.A. McAdams, C.L. Dym, Modeling and information in the design process, in *ASME 2004 International Design Engineering Technical Conferences and Computers and Information in Engineering Conference*, Jan 2004 (American Society of Mechanical Engineers), pp. 21–30

R.G. Quinn, Drexel's E4 program: a different professional experience for engineering students and faculty. J. Eng. Educ. **82**(4), 196–202 (1993)

K.A. Smith, P.K. Imbrie, *Teamwork and Project Management* (2004)

K.A. Smith, S.D. Sheppard, D.W. Johnson, R.T. Johnson, Pedagogies of engagement: classroom-based practices. J. Eng. Educ. **94**(1), 87–101 (2005)

J. Voogt, P. Fisser, J. Good, P. Mishra, A. Yadav, Computational thinking in compulsory education: towards an agenda for research and practice. Educ. Inform. Technol. **20**(4), 715–728 (2015)

K.L. Wood, D. Jensen, J. Bezdek, K.N. Otto, Reverse engineering and redesign: courses to incrementally and systematically teach design. J. Eng. Educ. **90**(3), 363–374 (2001)

K. Kumar et al., *Design Thinking to Digital Thinking*,
Manufacturing and Surface Engineering,
https://doi.org/10.1007/978-3-030-31359-3

Index

K. Kumar et al., *Design Thinking to Digital Thinking*,
Manufacturing and Surface Engineering,
https://doi.org/10.1007/978-3-030-31359-3

Made in the USA
Monee, IL
08 July 2022

ENVIRONMENTAL ENGINEERING DICTIONARY OF TECHNICAL TERMS AND PHRASES

ENVIRONMENTAL ENGINEERING DICTIONARY OF TECHNICAL TERMS AND PHRASES

English to Russian and Russian to English

FRANCIS J. HOPCROFT
AND SERGEY BOBROV

MOMENTUM PRESS, LLC, NEW YORK

Environmental Engineering Dictionary of Technical Terms and Phrases: English to Russian and Russian to English

First published by Momentum Press®, LLC
222 East 46th Street, New York, NY 10017
www.momentumpress.net

ISBN-13: 978-1-94561-238-1 (print)
ISBN-13: 978-1-94561-239-8 (e-book)

Momentum Press Environmental Engineering Collection

Collection ISSN: 2375-3625 (print)
Collection ISSN: 2375-3633 (electronic)

Cover and interior design by Exeter Premedia Services Private Ltd., Chennai, India

10 9 8 7 6 5 4 3 2 1

Printed in the United States of America

ABSTRACT

This reference manual provides a list of approximately 300 technical terms and phrases common to Environmental and Civil Engineering which non-English speakers often find difficult to understand in English. The manual provides the terms and phrases in alphabetical order, followed by a concise English definition, then a translation of the term in Russian and, finally, an interpretation or translation of the term or phrase in Russian. Following the Russian translations section, the columns are reversed and reordered alphabetically in Russian with the English term and translation following the Russian term or phrase. The objective is to provide a Technical Term Reference manual for non-English speaking students and engineers who are familiar with Russian, but uncomfortable with English and to provide a similar reference for English speaking students and engineers working in an area of the world where the Russian language predominates.

KEYWORDS

English to Russian translator, Russian to English translator, technical term translator, translator